# Risk Management in the Fire Service

# Risk Management in the Fire Service

Steven S. Wilder

PennWell Publishing Company

## disclaimer

The recommendations, advice, descriptions, and methods in this book are presented solely for educational purposes. The author and publisher assume no liability whatsoever for any loss or damage that results from the use of any of the material in this book. Use of the material in this book is solely at the risk of the user.



Published by Fire Engineering Books & Videos
A Division of PennWell Publishing Company
Park 80 West, Plaza 2
Saddle Brook, NJ 07663
United States of America

BOOK DESIGN BY CLIFFORD C. RUMPF
COVER AND ILLUSTRATIONS BY STEVE HETZEL

2 3 4 5 6 7 8 9 10
Printed in the United States of America

**Library of Congress Cataloging-in-Publication Data**

Wilder, Steven S. (Steven Scott), 1958-
Risk management in the fire service / Steven S. Wilder.
p. cm.
ISBN 0-912212-63-2
1. Fire departments—Management. 2. Risk management. 3. Fire departments—Law and legislation. I. Title.
TH9158.W55 1997
363.37'56—dc21 97-17792
CIP

# about the author

Steven S. Wilder is a 20-year veteran of the Bradley (IL) Fire Department. He currently holds the ranks of captain and director of training. In addition, he is a field instructor with the Fire Service Institute at the University of Illinois and is an instructor in the Bachelor of Arts in Fire Service Administration for Southern Illinois University, teaching courses in fire service risk management for both institutions.

Wilder is a partner in the consulting firm of Sorensen, Wilder, and Associates in Bradley, Illinois, which provides risk and safety management services to fire departments, law enforcement agencies, and EMS organizations. Active in risk management since 1983, he has also written and produced a fire department training video on the subject and has had his works published in numerous trade journals and newsletters.

# dedication

This book is dedicated to fire service training personnel around the world, who have made it their mission to train their students to the highest standards, and who constantly search for safer and better ways of slaying the beast.

# acknowledgments

Thanks to my many friends and associates for their help, encouragement, and support in making this and other projects possible. A very special thanks to:

Terry Sutphen of the University of Illinois Fire Service Institute for giving me the opportunity to expand into fire service risk management.

Stephen Frew, JD, Rockford, Illinois, for sharing his experiences, thoughts, wisdom, and friendship.

Chris Sorensen of Sorensen, Wilder, and Associates. I couldn't ask for a better friend or partner.

My parents, for showing me that I could do anything I put my mind to.

My fellow officers and firefighters of the Bradley Fire Department. Your professionalism shows in everything you do.

My children, Ryan, Daniel, and Megan. Being a young person in today's society isn't easy. Thanks for making being "Dad" the easiest and most rewarding thing in my life. I love you.

And most of all, to my wife, Debbie. I'll never understand what you see, but you have always believed in me and have always encouraged me to spread my wings and reach for the next level, never allowing me to become discouraged. Few men will ever find what I have found in you.

# table of contents

# preface

While the wording may differ from organization to organization, the mission statements of most fire departments usually focus on saving lives and protecting property in the safest and most efficient ways possible. Fulfilling this mission requires coordination of effort, cooperation throughout the heirarchy, prudent incident command, and a viable integration of all emergency services. Risk management practices are no less vital to the fulfillment of our mission.

Risk management is distinct in that it is a new idea in the fire service, even though it has been informally applied for years. Such activities are primarily concerned with safety management, loss prevention, and awareness of financial liability. Among the functions common to fire department risk management programs are the coordination of policies, management of programs and activities, effective use of committees and personnel, and centralization of related functions under a common umbrella. A strong network of communication is essential, as are education and vigilance among all fire service personnel. Ultimately, every member is an active player in his or her department's risk management program.

The role of any risk manager is as unique and varied as the organizational structures of fire departments throughout the nation and the world. No matter what title, rank, or position is held by that individual, he or she is responsible for a variety of loss prevention activities. The fire chief or chief administrative officer will often look to this individual for recommendations on issues of safety management; equipment procurement; compliance with recognized standards; and other areas where, if imprudently managed, serious losses could otherwise occur. At the same time, the department's risk manager must be an individual who can relate to and interact with officers and nonofficers at every level throughout the organization. Developing a new fire department risk management program can be a thankless, time-consuming, and exhausting job. Still, the resulting benefits to departmental operations far outweigh the challenges along the way.

# chapter one: risk management defined

Risk management in the fire service is a relatively new concept—one that most departments have discussed but which few have successfully implemented. In many agencies, the main reason for not inaugurating such a program is uncertainty. Officers and administrators are often hesitant to develop a professional risk management program because of questions about its purpose, its cost, and its appropriateness for smaller vs. larger departments. While a variety of questions might be posed, risk management must simply be viewed as a decision-making process designed to help fire departments determine where loss exposures exist, where unsafe acts or conditions can contribute to a loss, and how to financially deal with losses that cannot be avoided. To manage risk is to manage uncertainty—if we were able to predict when and where every loss would occur, we could easily take steps to prevent them from occurring at all. While such a scenario is unrealistic, it does serve to show that the concept of a risk management program is largely based on proactivity.

Just as organizational structures and operational activities vary from city to city, so too do the reasons fire departments must consider these programs. There are a number of reasons risk management is becoming more prevalent, including:

- an increasing number of job-related injuries to fire department personnel;
- an increasing number of injuries to the public resulting from fire department operations;
- the increasing accident rates of fire department operations;
- continually increasing insurance rates;
- the increasing frequency and severity of lawsuits against fire departments and personnel;
- increasing public awareness of fire department operations and a proportionate increase in expectations;
- questionable safety records; and
- poor preventive maintenance programs, resulting in increased downtime of apparatus and equipment, as well as equipment failures during times of emergencies.

To begin applying the concept of risk management to the fire department, it is first essential to establish a basic definition of the terms:

*Risk:* The possibility of injury or loss; the presence of a dangerous element or operational factor, known or unknown.

*Management:* The responsible supervision of an activity; the judicious use of means and resources to achieve a desired end.

Combining these two individual definitions, *risk management* can be defined as the responsible supervision of an activity, operation, or process so as to minimize the potential for loss, as well as to maximize safety for the involved individuals and teams.

Looking closely at this definition, it becomes obvious that such a program crosses all boundaries, integrating with all areas of fire department operations. It is *not* intended to interfere with normal operations; rather, it must be seen as complementary to them, serving to increase safety and protect the financial resources of the department and the community.

Note, too, that the decision to develop and implement such a program entails an ongoing commitment to a process. No professionally developed and organized fire department risk management program is self-sustaining. Accordingly, it cannot be a program developed to meet a specific goal on a one-time basis and then be discarded. To be successful, it must be accepted as a continuing process—one that will touch every aspect of fire department administration and operations.

## questions for discussion

(1) What steps has your department taken that might fall under the risk management umbrella?

(2) What has been done in your department to increase safety and minimize risks?

(3) Who is the person responsible for safety in your department? What does this person do? How can his or her program be improved?

(4) Thinking about your own department's loss history, what types of losses seem most prevalent? What has been done about them?

# chapter two: the nature of loss

In most departments, components of a risk management program are already in place. Modern innovations such as the use of firefighter accountability systems, PASS devices, and incident management are all part of a broader effort to reduce risk. Note that, by speaking in these terms, we are referring to much more than just fire suppression and rescue operations. In a well-run risk management program, the changes touch every division within the department. In fact, one of the key ingredients of successful risk management is the effective linking of all loss prevention activities into a single, unified program. Many departments have taken steps to increase the safety of firefighters performing emergency operations. This is particularly important since certain occurrences, such as a line-of-duty death, can result in a variety of losses, some of which can be quantified and some of which cannot. The quantifiable losses in an instance of line-of-duty death might include insurance claims and other monetary settlements, as well as adjustments in overtime schedules to keep the various shifts properly manned. The unquantifiable losses would include the emotional toll that such a tragedy exacts on the department as a whole. It is important that, as an industry, the fire service realizes that not all losses are measurable. All losses

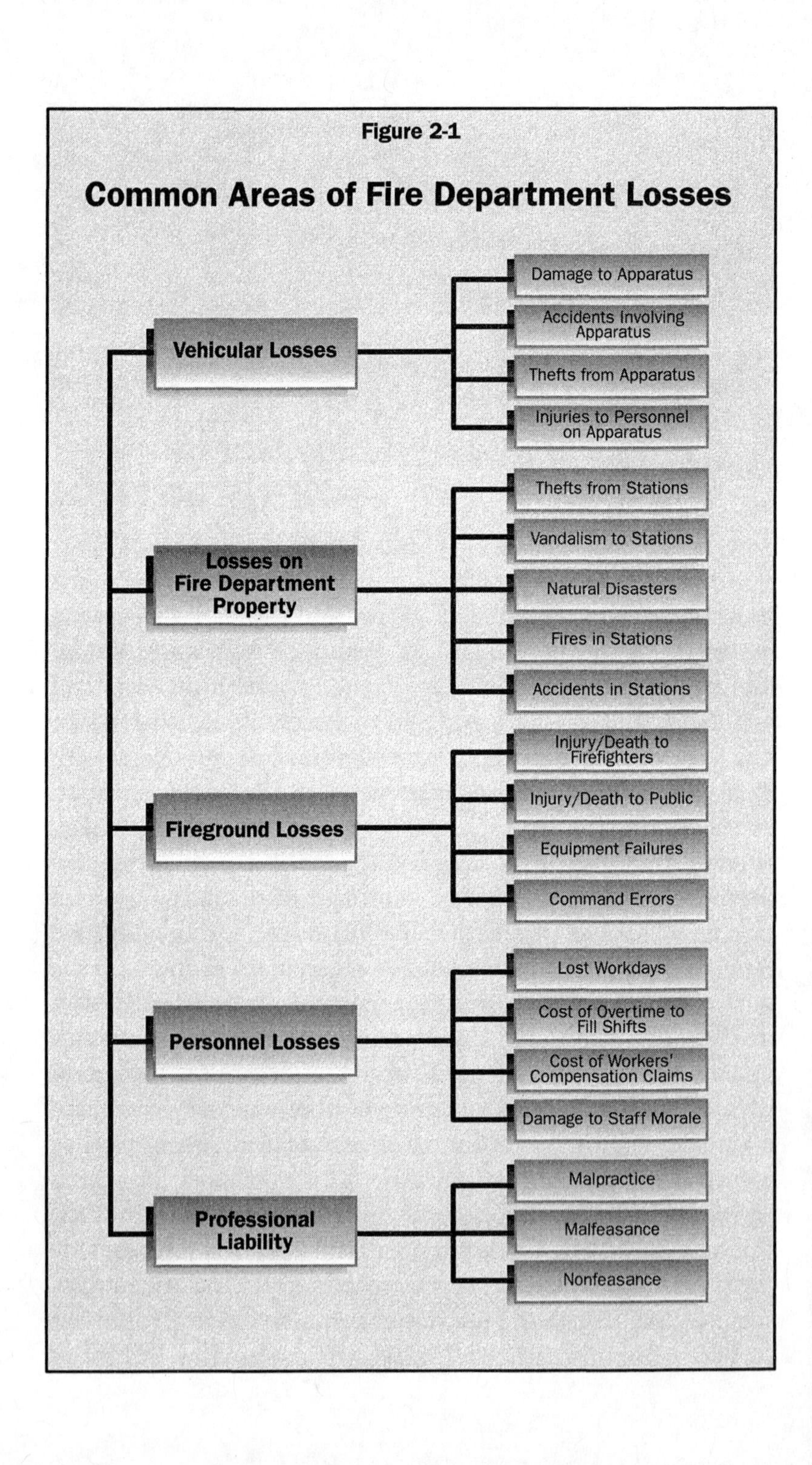
Figure 2-1
Common Areas of Fire Department Losses
Vehicular Losses
Damage to Apparatus
Accidents Involving Apparatus
Thefts from Apparatus
Injuries to Personnel on Apparatus
Losses on Fire Department Property
Thefts from Stations
Vandalism to Stations
Natural Disasters
Fires in Stations
Accidents in Stations
Fireground Losses
Injury/Death to Firefighters
Injury/Death to Public
Equipment Failures
Command Errors
Personnel Losses
Lost Workdays
Cost of Overtime to Fill Shifts
Cost of Workers' Compensation Claims
Damage to Staff Morale
Professional Liability
Malpractice
Malfeasance
Nonfeasance

weaken an organization and can lessen its efficacy in performing its duties.

In developing a fire department risk management program, proactivity is key to success. To be proactive, the program must try to identify potential losses before they occur. We can systematically categorize the common areas of fire department loss to more effectively predict them and plan the proactive steps that we must take to avoid them. Historically, losses in the fire service have fallen into five separate categories (Figure 2-1). They are:

- vehicular losses,
- losses occurring on fire department property,
- fireground losses,
- personnel losses, and
- issues of professional liability.

These losses, while listed in no significant order of importance or value, represent significant loss exposure to the department and deserve further review.

### vehicular losses

Losses involving fire department vehicles can have a variety of catastrophic repercussions. Among the most frequently occurring mishaps are motor vehicle accidents. Anytime a fire department apparatus is involved in a traffic accident, the damage that can accrue and the associated loss value can be overwhelming. A distinct correlation exists between accidents involving fire department apparatus and the risk of injuries, both to individuals in other vehicles and to the personnel on board the apparatus (Figure 2-2). Coupled with this is the likelihood of civil action from any and all parties involved.

If damage is significant, the fire department will also be faced with the loss of the apparatus, whether for the time of repair or replacement. Sometimes reserve vehicles are readily available; other departments don't have that luxury. When

Figure 2-2

## Fire Department Vehicle Accidents and Resulting Firefighter Injuries While Responding to or Returning from Emergencies in 1995

| | Number of Accidents | Number of Firefighter Injuries |
|---|---|---|
| Involving Fire Department Emergency Vehicles | 14,670 | 950 |
| Involving Firefighter Personal Vehicles | 1,690 | 190 |

**NOTE:** The National Fire Protection Association (NFPA) estimates that fire departments responded to more than 16.4 million incidents in 1995. Based on this estimate, the number of accidents represent about one-tenth of 1% of total responses.

Source: NFPA Survey of Fire Departments.

there is no substitute available, both the department and the community are denied a measure of protection, and some unquantifiable loss exposure results.

Equipment failures provide another example of potential loss involving apparatus. These failures can range from major mechanical breakdowns (including pump failure) to less obvious ones, such as dead batteries on EMS equipment. Regardless of the exact type or potential, the risk of loss associated with equipment failure is enormous. It accentuates the need for a well-designed and strictly enforced preventive maintenance program.

The theft of equipment from fire department apparatus is yet another aspect of vehicular loss. Sadly, most (if not all) communities have areas where increased criminal activity exists. In recent years, the problem has grown to include outright damage to vehicles and acts of violence against personnel. As a result, many departments have found themselves having to deal with the human element as well as the fire—essentially a war on two fronts. In such high-crime districts, it is incumbent on fire departments to take appropriate steps to protect their members as well as their equipment.

## losses on fire department property

Unfortunately, the risk of criminal activity doesn't stop at the fireground—it reaches back into the station as well. A quick survey of the contents of a typical fire station will reveal many items of potential interest to the criminal element. These range from narcotics and other controlled substances, carried on board advanced life support (ALS) ambulances, to forcible entry tools, highly prized by burglars. Because of our vulnerabilities, fire departments can no longer view their stations as sanctuaries but rather must take appropriate steps to provide the security the stations require.

Not all of the threats are posed by man, however. Natural forces can also cause their share of loss. Floods, earthquakes, hurricanes, and blizzards compromise a department's ability to provide service. They also result in losses to the department itself. Such losses can be severe enough to eliminate response capabilities altogether. While natural disasters cannot be avoided, fire departments must have contingency plans in place to deal with the risks.

One serious loss exposure that is often overlooked is the risk of fire in a fire station. As an industry, we pride ourselves on our ability to teach communities how to be fire-safe. At the same time, we often fail to observe the old adage that charity begins at home. A fire in a fire station can be devastating to the entire community, depriving everyone of protection. The principles of prevention that we teach must also be applied and practiced under our own roofs.

Fire departments must also recognize the risk of injury to personnel while they are in the station. Simple slips and falls, improper lifting techniques, and normal maintenance activities are three examples of situations that can lead to injury. Department administrators, risk managers, and safety officers must remember that the risk of injury exists not only at the scenes of emergencies but in the normal day-to-day activities as well.

## fireground losses

Among the most obvious and costly of losses that result from fireground operations are firefighter line-of-duty injuries and deaths. Such losses are both quantifiable and unquantifiable in nature. The calculable losses might include the cost of workers' compensation claims, overtime paid to fill vacancies, or even the cost of death benefits. Nonfinancial losses include the added risk of injuries to other firefighters if overtime is not paid or shift vacancies go unfilled, the increased risk to the community because fewer firefighters are available to provide the necessary services, and the psychological and emotional effects on other personnel within the organization.

Injury or death within the general public is another area of potential loss that you must consider. Anytime a fire department is on the scene of a true emergency, members of the general public gather. The fireground commander must be aware that civilians are present and ensure that adequate safety measures are maintained. If a member of the general public is injured while observing fire department operations, there is a strong likelihood that an issue of civil liability will arise. The fireground commander must be constantly vigilant and take appropriate steps to lessen or eliminate the risks.

Besides being detrimental to operations, equipment failure on the fireground can also lead to human loss. Many firefighters have become casualties as a result of equipment that failed during the course of emergencies. Pump failures during offensive attacks, burst hoses, and inoperational power tools have all been contributors to losses of this type. Because the risk to firefighters and civilians resulting from equipment failure is so great, it is incumbent on the fire department to develop and enforce an aggressive preventive maintenance program on all tools and equipment and to document whatever maintenance is done.

The final area of loss exposure to be discussed in conjunction with the fireground involves flawed command. Since the fire service has become more proficient in its use of incident management systems, this risk has significantly lessened. This is reflected in the accompanying statistics on injuries and deaths to firefighters in recent years (Figures 2-3, 2-4, and 2-5). Although the improvements we have made as an industry are commendable, it must be recognized that not all fire departments use an ICS or IMS. As a result, freelancing still occurs. Accepting the premise that freelancing kills firefighters, it becomes clear that incident management is vital to modern emergency operations, regardless of type.

While incident management has been proven to lessen the chances of loss on the fireground, it is also a system that, unfortunately, often receives lip service without being properly implemented. Many departments pay homage to the concept but fail to adhere to it during an emergency response. As a result, firefighters look to the commander for specific

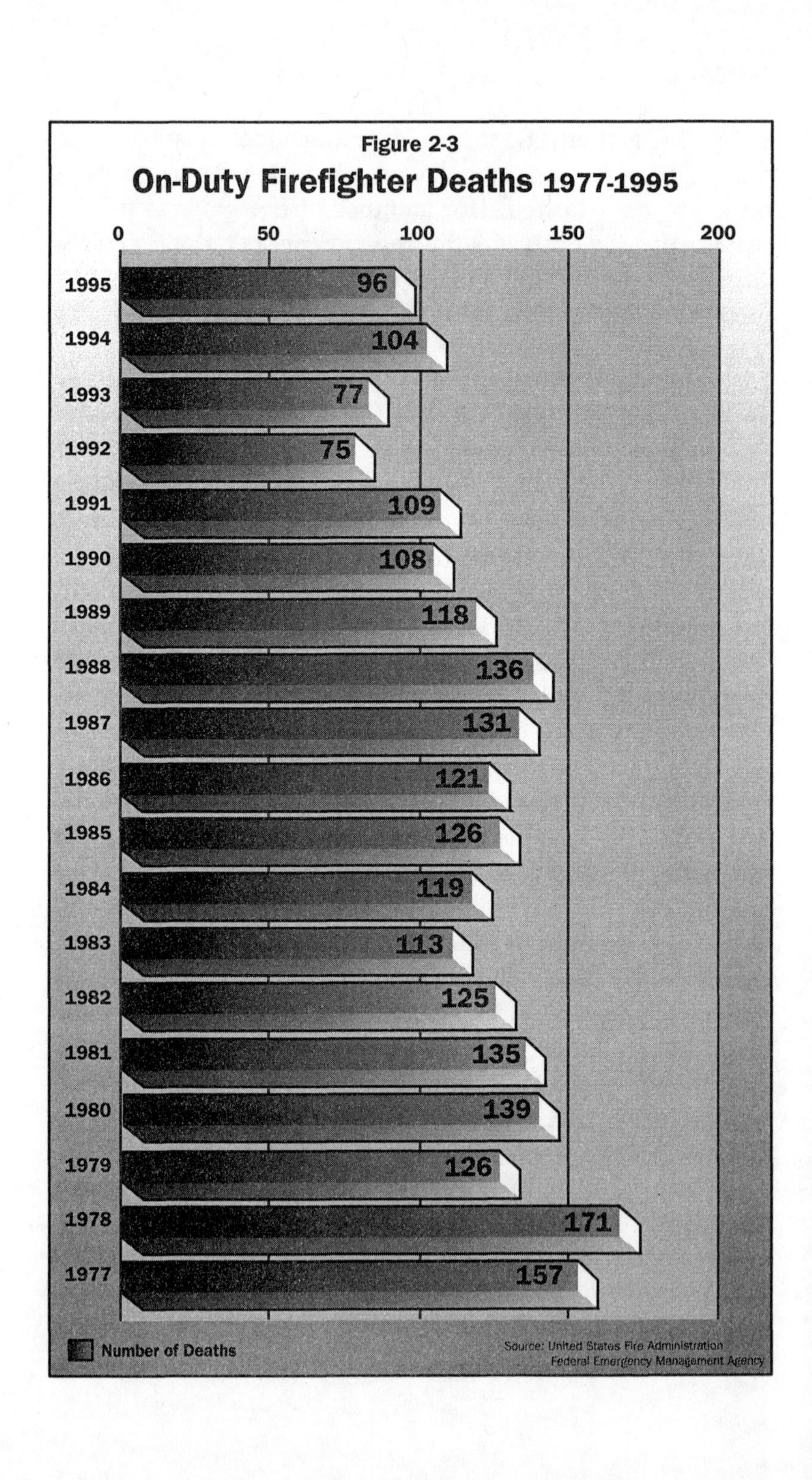
Figure 2-3
On-Duty Firefighter Deaths 1977-1995
0
50
100
150
200
1995
96
1994
104
1993
77
1992
75
1991
109
1990
108
1989
118
1988
136
1987
131
1986
121
1985
126
1984
119
1983
113
1982
125
1981
135
1980
139
1979
126
1978
171
1977
157
Number of Deaths
Source: United States Fire Administration
Federal Emergency Management Agency

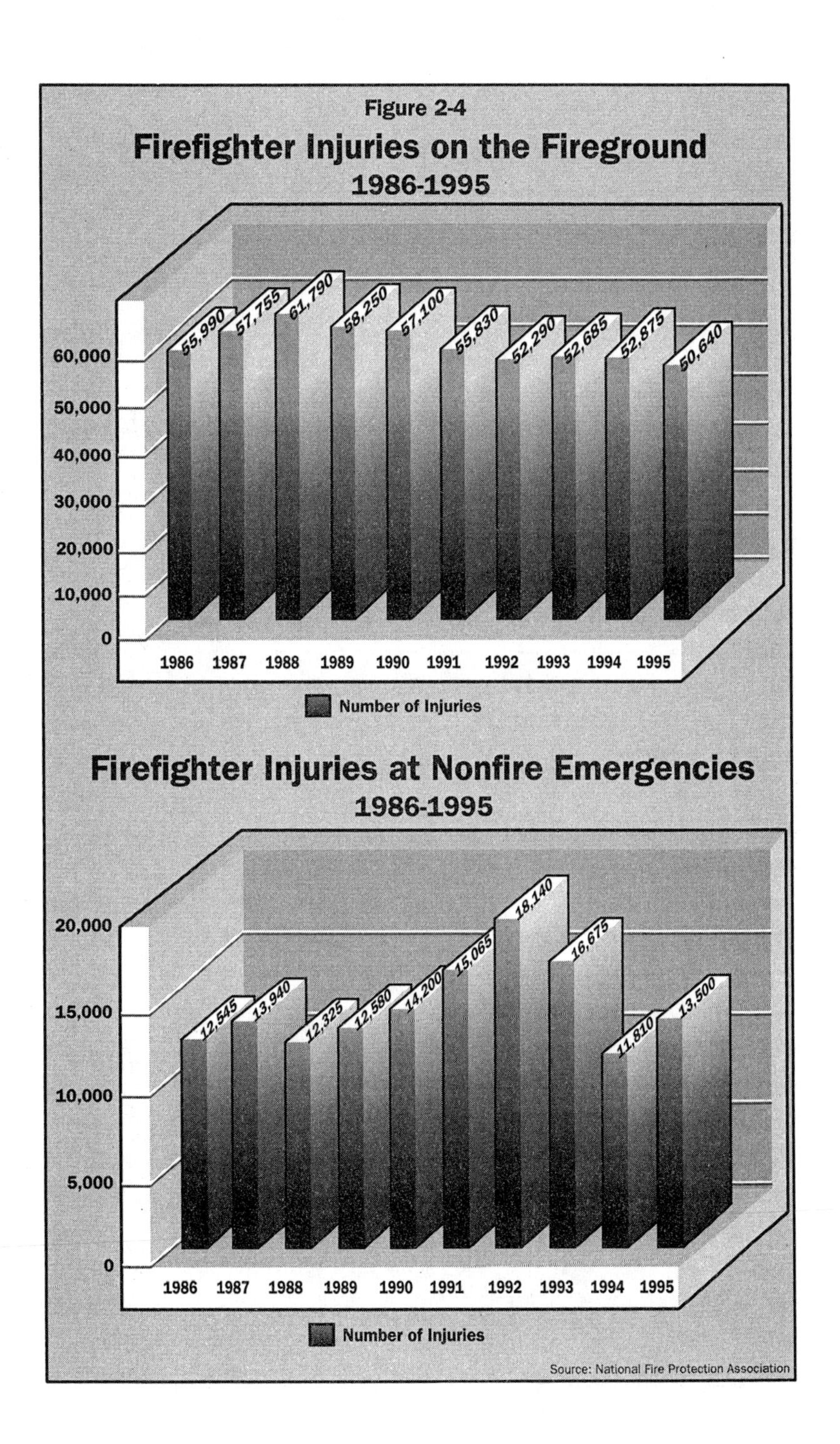
Figure 2-4
Firefighter Injuries on the Fireground
1986-1995
60,000
50,000
40,000
30,000
20,000
10,000
0
55,990
57,755
61,790
58,250
57,100
55,830
52,290
52,685
52,875
50,640
1986 1987 1988 1989 1990 1991 1992 1993 1994 1995
Number of Injuries
Firefighter Injuries at Nonfire Emergencies
1986-1995
20,000
15,000
10,000
5,000
0
12,545
13,940
12,325
12,580
14,200
15,065
18,140
16,675
11,810
13,500
1986 1987 1988 1989 1990 1991 1992 1993 1994 1995
Number of Injuries
Source: National Fire Protection Association

Figure 2-5

## Firefighter Injuries on the Fireground

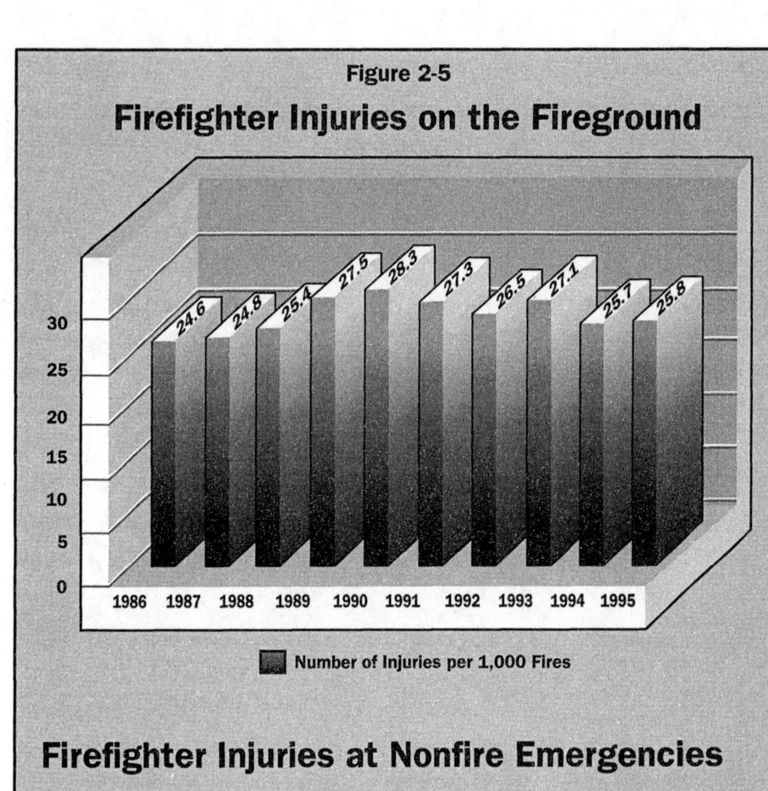


## Firefighter Injuries at Nonfire Emergencies

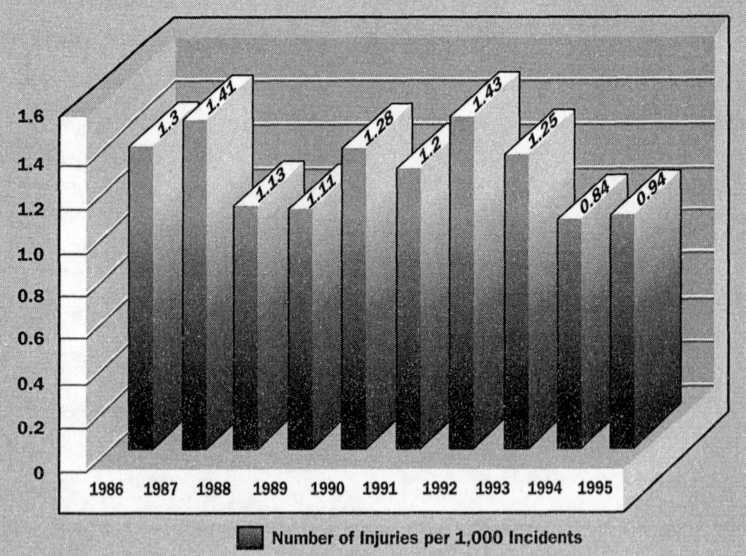


Source: National Fire Protection Association

instructions, only to realize that their commander is silently *encouraging* freelancing by his or her failure to implement the system. The result is often confusion among the firefighters and uncertainty as to who is responsible for what tasks—a situation that on occasion has had catastrophic consequences.

## personnel losses

Up to this point, we have looked at numerous scenarios that could potentially result in loss to the fire department. Often this potential has meant losses that directly affect personnel, whether in the form of workers' compensation costs, shift shortages, overtime, morale issues, or otherwise. Noting that one incident may result in losses over many areas is important. When developing a risk management program, the fire department administrator will realize that a proactive approach means loss prevention activity in a variety of areas.

No matter what the size or type of a given department, firefighting remains an industry focused on people. Apparatus, equipment, and buildings can all be replaced. The fireground commander must effectively employ a safety officer in all operations to maximize the safety of the fireground environment. Without doing so, personnel losses will continue to occur at alarming rates, and the associated losses will continue to accrue.

## professional liability

The issue of professional liability, as it relates to the fire department, focuses on performance. Throughout the fire service, universal standards have been developed and applied. While these standards aren't carved in stone as a model of performance, they *are* universally recognized as performance guidelines. Traditionally in the fire service, professional liability issues have broken down into three subcategories: malpractice, malfeasance, and nonfeasance.

Malpractice is a term commonly associated with medicine or law. Since the fire service is a primary provider of prehospital medical care, the issue of medical malpractice *does* deserve attention. As prehospital advanced life support services continue to progress, the issue of malpractice becomes more of a concern. Today, more drugs and advanced therapeutic practices are employed in the prehospital setting than ever before. Legislation governing EMS practices may vary from state to state, but all are based on certain standards of care. The fire department EMS provider is trained in these standards. Demonstrating proficiency in following them, he or she is then mandated to perform emergency care in the real world. Anytime a deviation from these standards occurs, and if the patient suffers a loss as a result, the potential for malpractice exists. Continued training, medical education, and continuous quality improvement activities remain the most proactive tools in lessening malpractice exposure. If for any reason the standards and protocols of the EMS system aren't followed, it should be on the orders of the emergency department physician. If this isn't possible, then it is incumbent on the EMS provider to document the type of deviation that occurred, as well as the basis for the decision. Since malpractice laws vary from state to state, it is up to the respective agencies to know the laws in their jurisdictions and to review the standards and protocols not only with their providers but with their project medical directors as well.

To understand the issues of malfeasance and nonfeasance as they relate to the fire service, you must first understand the concept of feasance. In the simplest terms, think of the term as meaning the "duty to act." Malfeasance suggests that the individual or organization had a duty to act but failed to act properly. Nonfeasance implies that the person failed to act at all. Lawsuits over issues of malfeasance and nonfeasance against firefighters and departments are rare; still, the potential exists. Again, the duty to act is based on standards, which in this case may include the department's own standards, established through the development of policies and procedures.

When developing policies and procedures, the fire department administrator is well advised to write them in such a way as to empower officers to make decisions rather than restrict them into performing in a specified way. Too often, policies and procedures in the fire service cause firefighters to perform in an exact manner. Such policies, while well intentioned, can be hazardous to an organization's operations, since they prohibit officers from making decisions based on circumstances. No two incidents are alike, and fire departments must have the flexibility to perform their duties in the most logical, sequential, and (most importantly) *safest* manner possible. This requires the fireground commander to size up each incident so as to make the appropriate decisions. Such decisions are made by qualified officers, not policies and procedures.

The fire service has been fortunate in that many competent fire service attorneys have published works with more detailed information on fire law. The modern department administrator should take advantage of the availability of such material and become acclimated to the laws and standards of the state in which he or she operates. It is also wise to have any department policies, procedures, and SOPs reviewed by legal counsel for appropriateness and vulnerabilities.

In this chapter, we have looked at a variety of areas where potential loss exposure exists. This isn't to suggest that there aren't more. As we will see in Chapter Three, each fire department must assess its own organizational culture and the specific type of service it provides. Many departments have expanded their roles into highly specialized areas such as confined-space rescue, haz-mat response, and urban search and rescue. Departments providing specialized services must look not only at the traditional services they provide but also at their innovations. This is essential toward identifying where loss exposures exist so that they can be proactively intercepted before they occur.

## questions for discussion

(1) Looking at your own department's history, does it seem as if the losses fit more into one category than any of the others?

(2) How are your policies written? Are they open-ended enough to allow your officers to make decisions, or do they cause firefighters to perform the same way at every incident?

# chapter three: the five-step risk management process

An effective risk management program is the result of a systematic approach. The five-step risk management process provides a framework for identifying and correcting loss exposures. The five steps of the process are:

(1) to identify and analyze, within the department's operations, areas where potential loss exposure exists;
(2) to develop alternative methods or processes that could lessen the loss;
(3) to choose the method that appears best suited to your department's needs;
(4) to implement the solution; and
(5) to follow up to determine outcomes.

In applying this process, it is important to realize that each one of these five steps, although performed individually, is interactive with the others. None of them will accomplish the goals of the program if they're not used in conjunction with the rest.

Figure 3-1

## The Five-Step Risk Management Process

| | | |
|---|---|---|
| Identify where loss exposures exist | **STEP 1**<br>Identify and analyze potential areas of loss exposure | Analyze the severity of the loss exposure |
| Safer ways of operating | **STEP 2**<br>Develop alternative methods to minimize the loss | Improved education and training at all levels |
| The department's current situation | **STEP 3**<br>Choose the solution that seems best-suited | Consider impact that changes will cause |
| Allow time for staff to adjust | **STEP 4**<br>Implement the chosen solution | Allow staff chance to question or comment |
| Periodic formalized follow-ups | **STEP 5**<br>Monitor outcomes | Measured against pre-determined outcomes |

To effectively use this process, it is also important to understand the applicability, as well as the application, of each.

## step 1:
### *identify and analyze areas where potential loss exposure exists*

In Chapter Two, we talked about the importance of taking a proactive approach toward managing risks in the fire service. Risks can be managed in two ways. First, they can be managed *reactively*, meaning that we can wait for a loss to occur, after which we go back and attempt to improve our operations so that such a loss won't be repeated. Or, they can be handled *proactively*, meaning that we try to identify potential losses in advance and act accordingly to prevent them from occurring at all. The first of our five steps is critical toward developing a proactive program.

In this step, we look at all of the activities that the fire department performs, plus the various operations within those activities. Since the services that your department provides may not be the same as those of neighboring departments, your assessment may be somewhat unique. Still, in most departments, we find common areas of operation. These include:

- administration,
- personnel,
- communications,
- in-station activities,
- public education,
- fire prevention and inspections,
- emergency vehicle operations,
- rescue,
- EMS operations,
- fire suppression,
- overhaul, and
- salvage.

This is a basic, composite list. Your own department may exclude some activities and include others. If your organization has haz-mat response teams, confined-space entry teams, or any other specialized services, then note those as well. Otherwise, your list will be incomplete, with the result that your program won't reach its full potential of proactivity.

Once you have listed the various activities your organization performs, break them down more exactingly, looking at each in terms of operations. In doing so, you'll begin to focus your attention not only on what it is that your department *does* but also where losses could occur. For example, let's look at the third item on our list, communications. Broadly defined, communications involves an array of equipment, a gamut of procedures, incoming calls for help, and outgoing dispatch instructions. Within this structure, where could events occur that would result in a loss? Could human error, failures, omissions, or equipment breakdowns result in a loss to the department? Some commonly identified exposures include:

- dispatching emergency response vehicles to the wrong address,
- the lack of a backup system to verify locations,
- the failure to maintain or obtain backup equipment in the case of technical difficulty, and
- responding companies being unable to communicate with each other or with the dispatcher due to faulty equipment.

Just in this simple exercise, we have identified four potential losses within one area of fire department operations. All four of these represent potentially catastrophic outcomes, both in terms of financial exposure and in the department's ability to continue to provide quality emergency services. Again, bear in mind that not all losses are quantifiable, nor are they all financial. If, as in the first example, an emergency vehicle were sent to the wrong address, how disastrous might be the outcome? Consider three scenarios:

• The responding equipment fails to arrive in a timely manner, resulting in loss of life, increased injury, or increased property loss.

• The responding equipment fails to find the address at all. Instead, it returns to the station, classifying the run as "unfounded." The caller never receives the needed assistance.

• The dispatcher is unable to recontact the caller for clarification and must therefore wait, hoping that the caller phones back a second time . . . soon.

In applying the first step of the risk management process, you'll find yourself looking at your own department and its operations through a different set of eyes. You will quickly realize vulnerabilities that you didn't know existed. You may also realize how fortunate you've been that you haven't already experienced a catastrophic loss. Your focus is different, and your intent is different. You may actually see things you may not want to see. At the same time, when you begin identifying all the various areas of loss exposure, you take a major step toward cutting losses, improving safety, and maximizing operations.

At this point, it would be to your advantage to devote a few minutes to listing all of the activities and operations in which your department is involved. Use the ones already mentioned as a foundation and build on them accordingly. Once you have done so, break them down into subcategories, identifying the loss exposures within each. This phase, whether performed individually or as a group activity, will require more than a few minutes. It will not be a quick process, nor will it be a simple process. On completion, however, you'll not only have a comprehensive list of potential risks within your department, you will also very likely have aroused the interest of others and have given them the opportunity to become involved in the development of your risk management program.

## other sources of information

Looking at your own department's activities is an excellent starting point toward identifying potential loss exposures. At the same time, other sources of information are available. These include:

*Your department's own past loss experience.*

While appropriate risk management documentation may not have been maintained, information should be available from your department's own past loss experiences. It can be gathered from former accident investigations, as well as past claims or suits (including information obtained during the discovery process, both from the plaintiff's and the defense's experts). Such information can help to point out where hidden vulnerabilities exist within your department.

*The loss experiences of neighboring departments.*

While no department likes to air its dirty laundry, the experiences of other local organizations can assist you in evaluating your own. If you know for a fact that a neighboring department has had experience in dealing with loss, a simple conversation with your counterpart can be worthwhile. Normally, this is best done *after* the case is closed and the organization has rebounded from its loss. By using another department's experiences as a learning tool, hopefully you can gain a clearer picture of your own department's situation.

*Historical data from insurance companies.*

Insurance companies exist for one primary purpose: to make money. In doing so, they are compelled to keep the cost of claims down. Approaching your insurance company with the intent of developing a risk management program should elicit a very positive response. Since only a limited number of companies insure fire departments, those that do should have sizable libraries of claims from which you can draw information and ideas. Also, most insurance companies have loss prevention personnel on staff who should be available to assist you. By working one-on-one, you'll find that many of the loss exposures that you're facing have already been experienced by others. As a result, methods of managing these risks should be available. Many of them have already been tested, with the results documented.

There are two points of caution to observe when dealing with insurance companies. First, they may want to charge you an hourly fee for working with you. This doesn't necessarily make sense, since you are developing a program that will ultimately save *them* money. Insist that they waive all hourly charges or any other consulting fees. If they refuse, ask them to negotiate a better rate. If they still refuse, it may be time to shop around for a new insurance carrier!

Second, when working with the loss prevention staff of an insurance company, insist that their representative be someone who is knowledgeable about the fire service. Many companies offer loss prevention "experts," but the fire service is a unique industry, deserving proper attention from qualified personnel. Few loss prevention specialists are able to apply their knowlege and skills to our operations effectively. While their representative may be very qualified, it is of no benefit to our risk management program to have to deal with someone who was at a shoe factory on Monday, an auto repair shop on Tuesday, a retail operation on Wednesday, a steel mill on Thursday, and the fire station on Friday. Demand the special expertise that our services warrant.

*Information obtained from legal counsel.*

Since your municipality or district already has corporate counsel on retainer, use the information he or she can provide. Most attorneys subscribe to some type of reporting service that keeps them abreast of changes in case law, notable outcomes, and damage awards. There are also specialized reports available that focus on legal issues and cases involving the fire service, and your attorney would have access to these. Ask him or her to forward this information to you on a regular basis. Review it to determine what parts, if any, might be applicable to your department.

*Information gathered from departmental incident reports.* Many departments have implemented incident reporting systems, in which an incident report is generated whenever there is an unexpected outcome, an injury to a firefighter or civilian, or a deviation from a prescribed process or standard. By reviewing these reports regularly, you can usually identify trends. Trends in occurrences often presage serious loss. When you identify one, apply the five-step risk management process and take appropriate steps to reverse the trend and eliminate the exposure.

## step 2:
### *develop alternative methods or processes that could lessen the loss*

In Step 1 of the five-step process, we identified areas where potential losses might occur within a typical organization. In Step 2, we'll look at the way we operate and seek better ways to reduce the potential loss exposure.

There are many ways to approach this, depending on your department's organization culture and the management style. Many departments have enjoyed success by using a brainstorming approach. Others have broken into small task forces, each one assigned a specific area of endeavor. Still others have decided to keep the responsibility for developing this second step at the administrative level. No matter which method is used, it remains a critical component in the process, for it is here that we truly begin to develop our proactive approach. Using our previous example of communications, our first identified risk exposure was dispatching emergency equipment to the wrong address. In Step 2, what might we consider to reduce or eliminate this risk altogether? Some possibilities:

- Keep the caller on the line until after the equipment is dispatched, then verify the address with the caller.
- Use an enhanced system that displays the caller's name, address, and phone number on the screen.
- Install a redundancy system that allows for playback of calls in case of errors or confusion.

For each of the potential loss exposures that you identify in Step 1 of the process, you should be able to develop two or three methods of dealing with it. On completion of this step, you'll not only have an extensive list of vulnerabilities, you'll also have an even more extensive list of ways to alleviate them.

## step 3:
### *choose the method that appears best suited to your department*

The third step begins the portion of the five-step process that requires strong administrative commitment and support. Here we direct our attention to the list of alternative methods that were developed in the second step, focusing on the ones that seem best suited to our needs. In identifying the best, it is important to consider not only the concept of loss prevention but also the development and implementation process as well. In addition, the fire service administrator cannot be oblivious to the fact that, in all likelihood, we are asking or expecting people to change, and change is not always quickly embraced.

Keep in mind that, in this third step, we are only *choosing* the solution that may be best suited. Our present intent is only to be able to give more consideration to it. Once you have chosen the solution that you think is best suited to your department's needs, you must examine it closely to identify what, if any, will be the consequences of pursuing it. After considering it from this perspective, you will know for certain whether this solution is the proper one and how to prepare your department for its implementation.

## step 4:
### *implement the solution*

Once you have chosen a solution, it is time to see how it might be implemented. Implementation can be a very challenging task, perhaps *the* most challenging in the process. Implementation involves change. Often, your staff may be resistant to change for a variety of reasons. For example, think back several years to when National Fire Protection Association (NFPA) standards for firefighter safety were first introduced. Do you recall the response when it became obvious to the fire service that three-quarter hip boots would no longer be acceptable and that bunker pants were going to be the standard against which they would measure us? At the time, many members of the fire service were upset over this forced change. Their resistance to it was based on a variety of reasons. Some simply didn't like bunker pants, others didn't like the idea that these changes were being forced on them, while still others objected to the cost that it imposed on the service. For whatever reasons, the fire service didn't receive these changes with open arms.

Unless they are properly implemented, changes within an organization that come about as a result of a risk management program can produce the same resistance. Within the overall concept of the program, introduce changes gradually, and give the people affected by the changes ample opportunity to pose questions, provide commentary, and offer suggestions. By doing so, you will lessen resistance, and the people whom the changes will most affect will have a much better understanding of the ultimate goal.

Another common mistake made by managers everywhere is in their propensity to use policies to dictate change. Policies in themselves should be seen as guidelines and be written as such. Many fire departments have rewritten their organizational policies and procedural manuals, thereby eliminating standard operating procedures. What they create instead are standard operating *guidelines*. Policies as such have traditionally been perceived and held as a means of action. When you act differently from the way the policy dictates, you deviate from your own standard. Deviation from a standard in the fire service may leave you vulnerable to serious legal liability.

In contrast to procedures, guidelines empower you to make decisions. They are designed to help you think and to make the right decisions. They aren't intended to tell you how to act, to tell you how to behave, or to make decisions for you in advance. To most administrators, writing policies becomes second nature. They have visions in mind of what they want to accomplish, and they use policies as road maps to turn their visions into reality. They write their policies, often in a step-by-step format, in ways that they believe are clear, concise, and which cannot possibly be misunderstood. Yet so often, the policies *are* misunderstood, and the outcomes are notably different from the ones intended.

Unfortunately, in the fire service, where human lives are constantly at stake, the margin for misinterpretation is small, and the outcomes can be catastrophic. To get a better understanding of how policies can be misconstrued, a simple exercise is often used in management training programs. It clearly illustrates how different people interpret the same policy.

*The Setup:*

This exercise requires at least three to four people to be effective. The rules are quite simple:

(1) No one can ask any questions.

(2) One person will serve as the moderator. He or she will read the step-by-step instructions given below.

(3) All participants, on the moderator's instructions, will close their eyes and keep them closed throughout the entire exercise. Only on the moderator's instructions may they open their eyes again.

*The Exercise:*

The moderator will give the participants the following instructions. He or she will not answer any questions.

(1) Take a sheet of blank paper out of your notebook. You won't need a pen—only a piece of paper.

(2) Hold the sheet of paper in your hands and close your eyes.

(3) Fold the sheet in half.

(4) Tear off the lower left-hand corner.

(5) Fold the sheet of paper in half again.

(6) Tear off the upper left-hand corner.

(7) Fold the sheet of paper in half again.

(8) Tear off the upper right-hand corner.

(9) Open your eyes and unfold your paper. Compare yours with those of the other participants.

Ordinarily, everyone who participates in the exercise winds up with a sheet of paper that looks different from all the others. This is expected. The instructions for the participants to follow were very clear . . . or were they? Were the folds in the paper to be made lengthwise or widthwise? Were the corners to be torn off the folds? Such ambiguities affect the outcome and might have led to appropriate questions had the moderator not been "managing by policy."

Similarly, the fire service administrator must be aware of the impact that any changes may have on his or her department and must make certain that the method of implementation is one that allows participation and input from those who will be most affected.

## step 5:
### *follow up to determine outcomes*

This is probably the most frequently overlooked step in the five-step risk management process, and for one simple reason: After preventive measures have been implemented in the fourth step, time must pass before their degree of success can be measured. Because of this time lapse, it is very easy to move on to other affairs and forget about this step altogether.

Initially, we found ourselves faced with a situation that presented an unacceptable level of risk. In applying the first four steps, we developed an idea of what we wanted to accomplish. In the fifth step of the process, we measure results against our original vision of the goals and question whether those results were what we had anticipated. If our outcomes match our goals, or are reasonably close, then we have probably averted loss or at least reduced the exposure to an acceptable level. If the outcomes are different from what we had projected, it is incumbent on us, the department, to identify the reason for the variance and to correct it.

One of the nicest features about this five-step process is its flexibility. When the outcomes don't meet our expectations, we can identify the reasons and make adjustments as necessary. If our chosen method turns out to be impractical, we can go back to the second step, analyze new possible solutions, and start the process all over again from that point. In this way, we can say that the five-steps are actually cyclical in design.

This five-step risk management process allows us to organize ourselves and our program so as to take a proactive approach toward lowering risks. Although it can be time-consuming when properly applied, it provides a strong foundation for our program, allowing us to develop better ways of managing our resources—translating, ultimately, into the best and safest service that we can provide.

**group activity**

(1) Using the information in this chapter as a guide, develop a list of all the operations and services that your department provides.

(2) Apply the first and second steps in the five-step risk management process. For each area listed above, identify five potential loss exposures. For each loss exposure, find three alternatives to reduce or eliminate the risk.

# chapter four: methods of managing risk

In Chapter Three, we introduced the concept of the five-step risk management process. The second step of this process, that of developing alternatives, is the one that allows us to look at the way we operate and to effect beneficial changes. Four common techniques can be used to develop these alternatives. They are:

(1) exposure avoidance,
(2) loss control,
(3) separation of exposures, and
(4) contractual transfers.

Each of these techniques has advantages and disadvantages, but all provide effective means toward lessening the loss exposure that the fire department may face during the course of its operations.

## exposure avoidance

Using the technique of exposure avoidance, the department simply avoids the loss exposure by dropping the risky activity. Let's assume, for example, that a fire department has been forced to deal with an increasing number of medical malpractice claims as a result of its EMS operations. Suits have been brought against the department and the paramedics, and the costs of defense, investigators, and discovery are escalating. As a result, the organization has to contend with sharply increasing insurance rates. All of these add up to a serious financial loss. In addition, the paramedics have expressed a fear of continued operation and are voicing their reluctance to provide prehospital services.

Using the technique of exposure avoidance, the department can avoid any future loss exposure by eliminating EMS services. Once these have been discontinued, the only remaining exposure is in legal actions that haven't yet been filed. Once any statutes of limitations expire in these cases, the exposure no longer exists.

Naturally, this option provides a very concrete outcome, in that the exposures are truly avoided. In reality, as it relates to the fire service, this approach is seldom practical. Usually the services we provide can't be found elsewhere in the community and may not be readily available through any other resource. Dropping activities may cause more harm to the community than the potential risks. This technique may have a detrimental effect within the department as well. Continuing with our example, the question of jobs and benefits has to be considered. If the department were to stop providing EMS, could it continue to justify the employment of paramedics? Would layoffs become necessary? Would the benefits of avoiding the risk, as opposed to managing it, truly justify the consequences to the organization as well as the community? Before even *considering* exposure avoidance, a department must sincerely attempt to find more effective means of managing the risks it faces.

### loss control

In using the technique of loss control, a department accepts the fact that the exposure will continue to exist. Unlike exposure avoidance, loss control truly permits department administrators to use risk management skills.

Loss control techniques allow a department to continue to provide a full range of services while doing so in the safest and most effective way possible. Although the risk of loss continues to exist, its presence is significantly reduced, and the severity of experienced losses are reduced as well. It is for this reason that loss control is the most commonly applied risk management technique.

Throughout the years, the number of firefighters killed and injured in motor vehicle accidents while on emergency calls has remained statistically significant and worthy of concern. Many departments have established minimum requirements that all drivers must meet before they are allowed to operate apparatus under any circumstances. In some departments, those requirements may be as slight as a proper driver's license and evaluation by a designated examiner; other times, it may mean completion of a certified emergency vehicle operator course, including proficiency demonstrations in a controlled setting, such as an obstacle course. Whenever an organization mandates specified training, it is employing the technique of loss control. Completing an emergency vehicle operators course doesn't guarantee that the apparatus won't be involved in an accident or that the driver won't make a mistake, but it does serve to increase the driver's awareness and skill, thereby lessening the odds of a mishap.

Another simple example of loss control technique deals with personnel riding on apparatus. Fifteen years ago, it was common to see firefighters standing on tailboards, hanging on to crossbars, while responding to emergencies. Today, safety considerations have curtailed that practice. NFPA standards have been written for both fire apparatus design and firefighter safety. Most departments now have established protocols that prohibit firefighters from riding on tailboards. As new vehicles continue to be designed with enclosed cabs, safety is also increased. By the same token, most organizations realize that having an enclosed cab isn't enough and so require that seat belts be worn by all personnel anytime the truck is in motion. These departments have recognized that, even in an enclosed cab, the risk of injury remains significant in the event of a traffic accident. This mandatory use of seat belts is another step taken to lessen the severity of loss. Tailboard-riding prohibitions, self-enclosed cabs, and seat belt requirements are all exemplary of the loss control technique. They are designed to reduce the severity of, rather than eliminate, the exposure.

### separation of exposures

Much like exposure avoidance, the technique of separation of losses is often a difficult one to apply to the fire service. To be effective, the department must have conducted a detailed analysis of all its operations not only in terms of the services it provides but also in identifying every possible exposure potential. In some industries, this isn't an exceptionally difficult task. In the fire service, it can often be impossible.

In a manufacturing plant, for instance, the processes are usually the same day in and day out. They are designed to meet production quotas and are repeated over and over. Proficiency is heightened accordingly, and plant managers can often challenge the production line personnel to increase their output without compromising safety. By the luxury of repetition, the loss exposures that may exist on the line are readily identifiable and thus more easily managed. When production changes are made, or when quotas are increased, the associated risk assessment is completed and risk management techniques are adjusted. Using the techniques that were previously established in conjunction with the new techniques developed as a result of the changes, plant supervisors can take new risk management steps. Using both the old techniques and new components can be considered a form of separation of exposure.

In the fire service, there is little repetition. Each response is unique, and each requires customized management. Seldom are two fires ever handled in the same way; rarely would the circumstances surrounding any two haz-mat incidents result in the same approach, and there is no such thing as a "routine" rescue. Since each emergency that the fire department responds to is unique, the concept of separation of exposures is difficult to apply.

To appreciate the concept of separation of losses as it relates to the fire service, imagine a department serving a community in which a set of railroad tracks divides the town. The community has only one fire station, which is located on the west side of the tracks. There are two routes available to get from the station to any location on the east side. The simplest and most direct route involves crossing the tracks at grade level, where flashers and gates preside. The second route, somewhat more circuitous, is by means of an overpass. When responding to an emergency on the east side, the department is forced to make a decision as to which road to use. If members choose the most direct means, they stand the risk of being delayed by a train. If they choose the overpass, this risk is eliminated, but their response time will be consequently and uniformly increased. Using the concept of separation of exposures, the first-responding company would take the shortest and most direct route, accepting the risk of a happenstance delay. The second-responding company would automatically use the overpass. Traveling by either road presents a risk of delayed response; however, by opting to separate, the department splits the loss exposures.

Obviously, this technique is one that may actually present as much risk as it attempts to manage. Because of the uniqueness of the fire service, the concept of separation of exposures is one that the administrator should understand, even though it may seldom be used.

## contractual transfers

The concept of contractual transfer is a commonly used risk management technique. By this method, the receiver of professional goods or services agrees to accept any actual or implied risks on the basis of a binding contract. This contract is often called an *indemnification agreement* or a *hold-harmless agreement*.

To apply this technique to the fire service, let's look at a hypothetical case. During Fire Prevention Week, it isn't uncommon for the fire department to play an active role in the elementary schools. Suppose that a fourth-grade class has asked to visit the fire station as a field trip. The teacher says that she would like to show the children where the station is, where the firefighters eat and sleep, and to show them around the trucks. By inference, she suggests that the children would really enjoy being able to climb on the apparatus. She tells you that the school will be sending home waivers for all of the parents to sign, giving their kids permission to go on the trip.

On the day of the outing, everything goes as planned. The students are given a tour of the fire station and are allowed to explore the trucks. As they are getting off, one child doesn't wait to be helped and instead jumps to the floor. He lands wrong, sustaining an ankle injury that requires medical treatment. Within a few weeks, the fire department receives a letter from an attorney representing the family of the injured child. The letter indicates that the fire department was negligent in watching over the children and that it failed to provide a safe environment for them, leading to the injury. The letter goes on to state that the family is filing a claim against the fire department for damages. Is the waiver that the parents signed a valid defense tool? Maybe yes, maybe no.

In some states, waivers aren't recognized, since some courts have ruled that asking a person to sign one is the equivalent of asking him or her to waive certain rights. If the department allows the incident to progress to this point, it has already waited too long.

Instead of allowing itself to be caught in a situation of liability, the department could have better positioned itself by using the concept of contractual transfer. When the original request for the field trip was made, an indemnification agreement could have been executed between the fire district or municipal body and the school district. Note that the agreement isn't between the *department* and the school district but between the *governing body* and the school district. Once executed, the agreement becomes a legally binding contract through which the school district agrees to indemnify the fire department from and against any and all claims or losses associated with the activity. In our scenario, when the claim is asserted against the department by the parents of the injured child, the department can turn to the school district to provide coverage on the basis of the executed contract. The school district then becomes the financially obligated entity.

This method of managing risks is a practical and viable one for fire departments with respect to social activities as opposed to emergency operations. Since it *is* a method of legal management, a department would be well advised to consult with legal counsel before attempting to use it. At the same time, departments should also determine the legal validity of waivers in the jurisdictions in which they operate. Finally, when entering into such an agreement with another organization, the fire department must be certain that the governing body of that organization is empowered to enter into such an agreement. Some insurance companies prohibit policy holders, such as the school board in our example, from entering into an indemnification agreement without prior consent from the insurance company. If this is a requirement, and the organization enters into such an agreement anyway, there may actually be no insurance coverage, which may in turn leave the fire district or the municipal body responsible.

There are as many different methods of managing risk as there are risks themselves. In most cases, fire departments find themselves employing loss control techniques more frequently than the other techniques discussed in this chapter. Considering the changing nature of the fire service, as well as the variety of challenges that firefighters encounter, loss control techniques remain the most practical. In addition, this approach allows us to manage not only our people but also the methods by which we deliver the services we provide.

## questions for discussion

Examine the loss prevention steps that your department has taken so far and categorize them according to the four types that we have discussed:

(1) exposure avoidance,
(2) loss control,
(3) separation of exposures, and
(4) contractual avoidance.

Can you identify the method that has been used more frequently than the others? Which one? Why do you think this is so?

# chapter five: the safety committee

Next to the support, direction, and commitment given by administration, an effective safety committee can be one of the best tools for the success of a risk management program. If the safety committee is properly structured, its input and guidance can give a department ideas, objectives, and direction.

While most departments have formed such committees, their success seems to vary. Among the reasons for this are differing levels of education and experience among the members, degrees of commitment of the administration toward the safety committee, the leadership of the committee, variable follow-through in committee activities, and the relative seriousness with which personnel in the department adopt the recommendations. While a strong and functional safety committee can serve as the backbone of a risk management program, a dysfunctional one, or one that fails to provide leadership and direction, can be the basis for the demise of the entire risk management program. Often, the success or failure of the program rests in the hands of the safety committee leadership.

The safety committee chairman should be an individual who has received additional education and training in safety management practices. He or she must possess a variety of traits and qualities to be successful and must be able to communicate the importance of safety management to other members of the department. Additionally, this individual must be able to motivate people, understand them, and have a knack for accomplishing tasks by delegating them successfully. Too often, the safety committee is perceived as an entity whose responsibilities reside in its chairman, yet this is not the case. The chairman facilitates the activities of the committee, providing leadership and direction. Successful outcomes are based on the activities of the committee as a whole, and the responsibilities and accountability must be shared throughout.

While fire department safety standards are applied universally, they must be seen as minimum guidelines, not minimum outcomes. Recognizing that the organizational culture of each department is different, the responsibilities of the safety committee will vary. In some departments, safety committee activities may include research, program and policy development, implementation of safety programs, and authority in enforcing all occupational safety and health standards. In other organizations, the committee may be organized to provide direction, suggestions, and recommendations to administration. Regardless of the structure of the department, or of the particular responsibilities assigned, authority must be given to the safety committee if it is to fulfill its charter.

## safety committee membership

The composition of the safety committee should represent key areas of department operations, specialized services, and general membership. The specific organizational structure will depend on the committee's goals and objectives. In many departments, the committee membership is designed to include:

- the safety committee chairman, who is appointed by the fire chief or the department administrator;

- an administrative representative;
- a union representative, if applicable;
- representatives from specialized services, such as haz-mat teams, confined-space rescue teams, USAR teams, and the like;
- a training division representative;
- a member or members at large; and
- others as deemed appropriate.

When individuals are assigned to the safety committee, the responsibilities of that committee should be made clear to them. Obviously, this requires that the program itself be clearly defined, including an accurate description of the members' roles. Individuals appointed to the committee should also receive additional training in safety management practices so that their contributions are consistent with the overall goals.

The goals of the safety committee should also be clearly defined; they should be quantifiable and reviewed and updated no less than annually. Without clear objectives, the committee can lose focus easily, and successes that should be enjoyed can be missed. When that happens, the safety committee is no more than a social committee.

## establishing goals

In establishing the goals of the committee, the chairman, in concert with administration or the administrative representative, should identify what the expected outcomes are and how the committee can best serve the needs of the department. As a component of the risk management program, the safety committee can use risk assessment to indicate where it might make its contribution as well as where its leadership and expertise can best be used.

As previously stated, the objectives of the safety committee must be quantifiable. The objectives' outcomes must be measureable so that it is possible to know whether they have actually been fulfilled. One objective of the safety committee might read as follows:

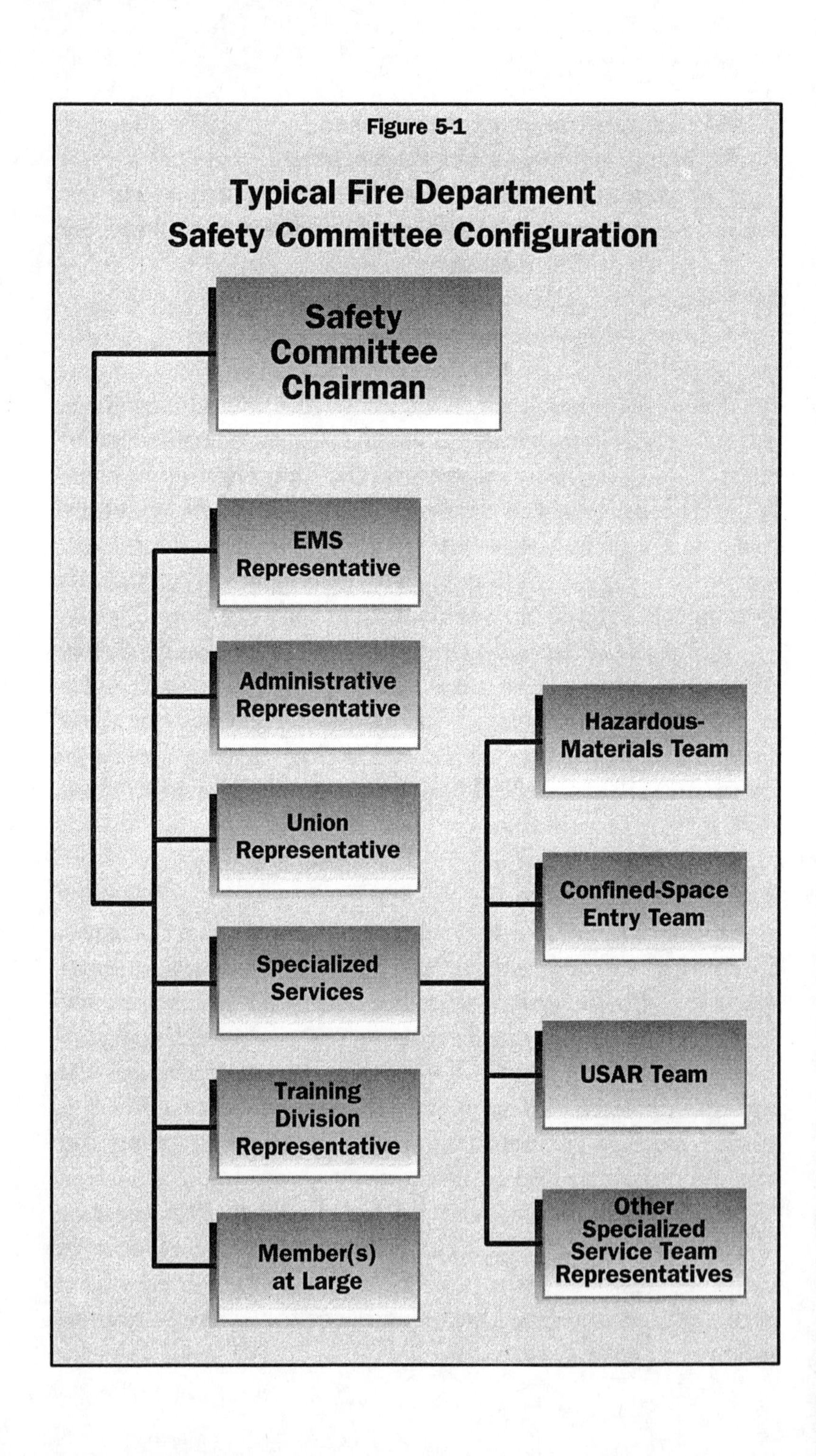
Figure 5-1
Typical Fire Department
Safety Committee Configuration
Safety Committee Chairman
EMS Representative
Administrative Representative
Union Representative
Specialized Services
Training Division Representative
Member(s) at Large
Hazardous-Materials Team
Confined-Space Entry Team
USAR Team
Other Specialized Service Team Representatives

*"The safety committee will conduct an assessment of all job-related injuries occurring in the fire stations over the previous 24 months. It will identify trends and common causes and provide a list of recommendations to the administration on ways to reduce or eliminate these types of injuries."*

Unfortunately, safety committees will often develop objectives that are ambiguous, that can neither be measured nor assessed. For example, how could this safety committee objective be measured?

*"The safety committee will review all fire department operations and take steps to eliminate injuries."*

While this objective is noble, it is irredeemingly vague. Its outcomes can't be measured; consequently, neither can its success. More often than not, we see objectives written in such a manner. As a result, the members of those committees find themselves uncertain as to what they are supposed to be doing and may even be unclear as to the mission of the safety committee.

It is up to each department to determine exactly who in the heirarchy is responsible for defining the goals of the safety committee. Regardless of whose task it becomes, that person or group must have a firm idea of how the results of the committee's work can be measured.

## the safety committee as part of the risk management program

Commonly, the safety committee will serve multiple functions within the risk management program. Since safety assessments are usually assigned to it, the committee becomes a key source of information toward completing Step 1 of the five-step process. You'll recall that the risk management process is ongoing, often cyclical. Regularly scheduled safety assessments provide a good example of how the process repeats itself. Each time a safety assessment is completed, the first step in the five-step process has been repeated. Consequently, new information becomes available through which the risk management program can be improved. Every time we identify a way to improve the program, we effectively lessen the risk of loss in any of the several categories mentioned earlier.

The department may also rely on the safety committee to provide recommendations on ways to improve the program. Obviously, this correlates with Step 2 of the five-step process, in which we identify methods of operation that will lessen loss exposure. For instance, suppose that after assessing all job-related injuries in the fire station over the past 24 months, the most common trend turned out to be slips and falls on wet floors that had just been mopped. What are some of the steps that the department might take to lessen this risk? Appropriately placed signs might be one answer. A general announcement could be made in the station that the floor in a given area will be wet and that personnel should avoid walking on it until further notice. Perhaps a rope or tape could be used to cordon off the area, or mopping could be done at other times when there is less activity in the station. Though simple in nature, any one of these solutions might prove to be effective, and the department could experience a significant reduction in such accidents. By making positive recommendations, the safety committee fulfills the second step of the five-step process.

## safety committee visibility

To be viable within the fire organization, safety committees must remain visible. When they are not, they are often forgotten, and their efforts aren't taken seriously. In a sense, they become token committees.

Increased visibility can be achieved in a number of ways. In some departments, when the safety committee meets, other members of the organization are invited to observe. Before adjournment, any members present have the opportunity to ask questions or make suggestions. By doing so, everyone can be a part of the process, and the resulting recommendations might not seem so alien to them. One word of caution when considering this approach: Realize that, in larger departments, the attendance could become so enormous that the open-forum segment might easily deteriorate into a scene of debates and verbal brawls, which will only serve to undermine the better efforts of the committee.

Another option that has proved successful is to distribute printed documentation of the committee's meetings and activities. Since minutes of all safety committee meetings should be duly recorded and maintained, they will be readily available for copying. Whenever the minutes are posted, employees have the opportunity to review them, discuss them, ask questions of the committee members, and provide their own suggestions. The department can take this one step further and post notices of the committee's meeting schedule, along with the agenda. Employees can then be asked to give their comments and ideas to the members before the meeting. In preparing the agenda, the safety committee should designate sufficient time so that all of the employees' submissions can be aired and discussed. While this is an effective way to solicit input, the employees must realize that submitting an idea doesn't necessarily mean that action will be taken. It still remains up to the safety committee to decide on any information after carefully weighing its pros and cons.

## safety subcommittees

As the safety committee grows, and as its functions and activities increase, it will likely reach a point where it won't be able to review all of its slated business in just one meeting. To avoid this overload, many have developed subcommittees that receive information from the main committee, meet independently from it, and then return their recommendations for final action or approval.

If having such subcommittees is deemed beneficial to the department, then it must be determined exactly what types are needed and how they should be structured. Again, this is a decision that will vary according to the specific nature of the organization. Certain subcommittees, however, seem indigenous to all departments:

- the firefighter health and safety subcommittee,
- the apparatus and equipment subcommittee,
- the training and education subcommittee, and
- the EMS subcommittee.

*The firefighter health and safety subcommittee.*
This subcommittee is commonly charged with a variety of responsibilities related to its overall theme. It is within this group that all reports of firefighter deaths and injuries are reviewed, including all unsafe acts or conditions leading up to the losses. This subcommittee should report its findings to the entire safety committee so that appropriate remedial action can be taken.

The firefighter health and safety subcommittee is also the entity responsible for monitoring all occupational safety and health requirements within the department itself. This includes hepatitis immunizations, annual physicals, and the like.

*The apparatus and equipment subcommittee.*
The apparatus and equipment subcommittee is the group that reviews changes in safety standards related to vehicles, tools, and equipment and makes recommendations as to how the department ought best stay in compliance with the standards. This may include updates in vehicle design, changes in turnout gear, safety deficits in existing tools and equipment, and information on the development of new equipment. This subcommittee should work closely with other entities within the safety committee, since many of their concerns overlap.

*The training and education subcommittee.*
This subcommittee is responsible for assessing the department's training program to the extent that safety measures are

incorporated. In addition, this subcommittee receives information from other entities about areas in which training and education must be provided. For example, if the apparatus and equipment subcommittee recommends acquiring a new type of power saw, and if the safety committee and administration agree, then the training and education subcommittee would work with the training division to ensure that proper training is given to personnel. "Proper training," of course, entails attention to safety considerations as well as general operation, and it should be given before the saw is ever put into service.

This subcommittee should also receive information from the training division about planned safety training for any given fiscal period. In reviewing this information, the training and education subcommittee reviews the training syllabus, then sends the training division recommendations on incorporating changing safety standards into existing programs. In doing so, the subcommittee and the safety committee as a whole become active participants in the department's ongoing training and education programs.

*The EMS subcommittee.*

Since emergency medical services cross into all of the other areas we've identified, this subcommittee actively participates with the rest. Its members should be individuals who are fully knowledgeable about both EMS and firefighting operations. Many of the requirements for EMS continuing education, equipment, and apparatus design fall under standards that are different from those that govern firefighting. Therefore, a strong *operational* link exists between the EMS subcommittee and other subcommittees as well. This entity works with the training and education subcommittee to ensure that the EMS training programs incorporate all of the latest safety standards and requirements. In the same way, they work with the apparatus and equipment subcommittee to ensure that new purchases for EMS use meet the latest standards for patient and personnel safety. The success of an EMS subcommittee is based on an appropriate mix of both EMS and non-EMS members and a willingness of all to work openly and in conjunction with the other groups.

The safety committee ultimately becomes one of the key components of any fire department's risk management program. A strong, functional, outcome-focused committee sets the stage for the results that we desire. Accordingly, a weak safety committee, with no direction and poor leadership, can nullify all of our other efforts and may even take us in a direction opposite to the one we wanted our program to take.

It is up to the fire chief or the administrator to stay informed of the progress of the safety committee and to compare the outcomes with the stated objectives. If the committee is succeeding in its task, then the chief or administrator should continue to offer the requisite support and resources. If the committee is faltering, then the chief or administrator must work directly with the chairman to identify the causes and develop a plan for getting back on track.

## questions for discussion

(1) Does your department have a *functional* safety committee? If not, what improvements might be made?

(2) What are the goals of your safety committee? Are the outcomes measured against them to assess their efficacy?

(3) What have been some of the major accomplishments of your safety committee over the past two years? How has information about these accomplishments been shared with the rest of the department?

(4) If asked, would all personnel within the department be able to identify improvements that have resulted from the efforts of the safety committee?

# chapter six: incident investigation

One of the most unfortunate situations that a fire department can find itself in occurs when an activity or event results in a claim against the department. Even under the best of circumstances, a claim can be a time-consuming and costly affair, as well as a scar on the department in the eyes of the community. No matter the size of the organization or the services it provides, there will be times when events can result in a claim. The department must treat such circumstances in the same proactive manner as it would any other element of its risk management program.

On experiencing a potentially compensable event (PCE), the focus of the department's response is twofold. First, the department must investigate the occurrence in a timely manner and determine the circumstances that led to it. Second, the department must realize that a PCE can become a visible incident, meaning that it may be played out in the media and become a point of community concern. This is important not only in terms of public relations but for its financial repercussions as well. From both points of view, a thorough investigation is critical toward making the right decisions and acting in ways that are both fair and reasonable.

## step 1:
### *ascertaining the facts*

For discussion, assume that an ambulance is dispatched to a residence to transport a person to the local hospital for a mental health evaluation. The dispatcher advises the crew that the patient is a 56-year-old male with a history of mental illness. The caller had described him as being belligerent and combative. Following department protocol, the dispatcher sends an engine company to assist the ambulance. On arrival, the patient's wife meets the responding crews at the front door. She advises the paramedics that the patient has a history of paranoid schizophrenia. She tells them that he is on two different medications but that he stopped taking them several days ago. Because the patient threw away the bottles, the wife is unable to identify what the medications were. She says that he has now locked himself in the bathroom and is threatening to harm anyone who attempts to approach him. She further states that she has spoken to his physician, who has ordered him brought to the emergency room of the local hospital for an evaluation. The wife also says that there are no guns in the house and that the patient is unarmed.

The paramedics position themselves in the hallway on either side of the bathroom door and identify themselves. After a lengthy discussion, the patient cooperatively exits the bathroom and walks out into the living room. The paramedics and the members of the engine company are all present. When the ambulance cot is brought in, the patient suddenly experiences a mood change and begins to threaten the personnel. All efforts to control the situation fail, and it becomes necessary to use force to get the patient onto the cot. During the ensuing struggle, and while being restrained on the floor, the patient ceases to resist. The paramedics and firefighters relax for a moment and suddenly realize that he isn't breathing. Resuscitative equipment is brought in, and the patient is coded. All of the resuscitative measures fail, and the patient is pronounced dead in the emergency room approximately 45 minutes later.

This hypothetical case presents several questions about liability exposure and compensability. Because of the untoward outcome, the incident must be considered a PCE, thus warranting an investigation. The department will use the information obtained from the investigation in a couple of different ways. First, it will be used to evaluate operations and to identify any existing weaknesses or vulnerabilities, later to make changes. Second, if a suit is filed against the department, which must be anticipated in such a case, then the findings of the investigation will be valuable to both the insurance company and the defense counsel.

In conducting the investigation, it is important to realize at the outset that the perspective of every individual involved may be different. The same will hold true of the patient's family members. Respecting the perspective and opinion of each person is important. The purpose of the investigation is to make *factual* determinations of what actually happened, not to draw conclusions. This is only done by separating the facts from the emotions, impressions, and opinions. Although some may argue against conducting a formalized investigation, doing so can be useful toward assessing the full details of the event. Another practical rationale for conducting a timely investigation is that the incident will still be fresh in people's minds; thus, more details will be obtainable.

From these discussions and interviews, it should be possible to assess the feelings, behaviors, and attitudes of everyone involved. This can be a critical step, since it allows the department to provide information to personnel who might need it. Properly documented, such information can save the department money in claims management costs and will be especially useful to defense counsel if a claim is ever filed. It must be stressed again that the purpose of the investigation is to gather facts, not place blame. If a claim is ultimately filed against the department or against any of the individuals, a unified defense is vital. If department personnel feel that the investigation was conducted for the purpose of placing blame, then everyone will become defensive. The result is finger pointing between the parties. There is no quicker way for a unified defense to break down than when parties within the defense team, including the named defendants, start blaming each other.

If the administration of the department doesn't feel comfortable investigating the incident, or when legal counsel advises against a self-investigation, then an outside investigator may be used. Often the department's insurer can provide a claims investigator for this purpose. Additionally, the department must be aware of laws regarding discovery within its jurisdiction. In some states, these laws may allow the department to protect documentation of its investigation from discovery and admissibility in court. In other states, there may be no such protection, and any records that exist become admissible evidence. If you are unsure about the laws in your area, you are well advised to consult your insurance carrier or legal counsel before beginning any investigation.

## step 2:
### *securing records and documentation*

If, after initial discussions and interviews, the department determines that there is a risk of a claim, it should flag and secure all records and documents pertaining to the event. The files should be maintained in a secure mode until resolution of the case or until any statutes of limitations have passed. Some departments have actually turned the original records over to their insurance carriers or legal counsel.

Using the scenario given above, should the family members request copies of the department's records, or should a request be received from their legal counsel, then the department should contact its own attorneys before releasing any documents. In most areas, a department is allowed reasonable time to produce documents after a request has been received. This period allows the department to review the records with its own legal counsel to determine whether they should be produced. One golden rule that the department ought to remember is, *Always let attorneys talk with attorneys*. If a request for records is received, it is incumbent on the department to let its attorney deal directly with the family's attorney. In cases where the department's insurance carrier assigns counsel, then he or she should be the one who deals with the family's attorney. Under no circumstances should a fire chief, department administrator, safety officer, or other member of the department attempt to deal directly with the lawyer representing the family. This type of law is highly specialized, and the department's best interests can be served only when discussions and negotiations are performed by a professional entity.

The critical nature of this phase cannot be understated. This is especially true when the department finds itself working with insurance carriers or legal counsel. In such cases, the carrier or its counsel has to feel confident that the department is providing all of the necessary information, no matter how sensitive, embarrassing, or confidential it might be. Also, the department has to feel confident that the insurance carrier will represent the department's best interests and that any legal counsel assigned by the insurance carrier will be both competent and dedicated.

## step 3:
### *the follow-up*

Once the investigation is complete and the department or the insurance carrier has determined that legal action is likely, several things can be done to minimize the potential damage. Keep in mind that any costs incurred in defending the case will be charged against the department's insurance policy and that these costs will probably affect future premiums or even the organization's insurability.

During both the initial investigation and any follow-up or discovery activities, the department should inform the members who were directly involved about the status of the case. These members should be reminded of the need for absolute confidentiality and not to discuss the case with anyone. This is to protect the department and its personnel in case any unauthorized or inappropriate contacts are attempted. The involved employees should also be kept posted on the status of any developments relevant to the incident and be reassured that all efforts are being made to insulate them against individual risk as well. It is critical that these involved members not feel as if they are forgotten or expendable.

In conclusion, fire departments must be aware of the risk of legal action in today's litigious society. While the risk management program is set up to prevent losses resulting from incidents that may lead to claims, situations arise that could result in losses. A good risk management program provides the opportunity to reduce the risk of a claim by identifying such incidents at the time they occur, conducting a timely investigation, and documenting all findings.

None of this is to suggest that prompt intervention won't prevent a claim from being filed sometime in the future. Still, in those instances when a potentially compensable event occurs, action taken by the department prior to a claim being filed can help reduce the lost exposure through time savings, cost savings, and prudent claims management. Accepting the fact that losses will occur and acting appropriately to minimize the damage of a PCE are two more components of a good fire department risk management program.

## questions for discussion

(1) Does your department have an established process for investigating those incidents that may result in legal action?

(2) Does your department have the appropriate forms available to document unusual activities or outcomes at the time of an occurrence?

(3) Has anyone from your insurance company or legal counsel spent any time educating your officers about their role in investigating potentially compensable events?

(4) Does your department have a policy regarding confidentiality?

# chapter seven: managing risks and emergency operations

So far, we have looked at many things a department can do to develop a proactive risk management program. Many of these have involved policies and guidelines, a review of existing SOPs or SOGs, and discussions on ways to maximize operational safety. These administrative measures are, in essence, the backbone of our program. Usually they allow us to effect changes to prevent losses under controlled circumstances.

During emergency operations, circumstances may not be as easily controlled. When responding to or operating at an incident, the ability to manage risks is significantly compromised. Unlike managing risks in a nonemergency situation, dealing with them at an incident is exceedingly complex, often due to changing conditions. Managing risk is just one of the many concerns the incident commander faces. Considering the number of variables to be handled, the chances of unsafe acts or conditions presenting themselves increase dramatically. No matter what the size of the emergency, do not overlook the principles of risk management.

## the incident command system as a risk management tool

Over the years, many different styles of the incident command system have been developed. All of them have incorporated various features of risk management, even while not necessarily calling it by that name. Most of these systems, for example, advocate the assignment of a scene safety officer, qualified by education and training, regardless of the type of incident. The scene safety officer in a sense becomes the scene risk manager. It is that person's responsibility to monitor changing conditions. When a high-risk situation is identified, he or she intervenes accordingly to prevent a loss from occurring. While the role of the scene safety officer is frequently equated with managing the safety of *personnel*, he or she can also play a critical role in preventing losses to those who are affected by the incident, such as property owners, occupants, and bystanders.

All of the various models of incident command recommend that the scene safety officer be empowered to intercede whenever the safety of personnel is jeopardized. This power, not to be taken lightly, shows the importance that risk management must be accorded at emergencies. In Chapter Five, we discussed safety management, noting that every incident that results in a loss is caused by either an unsafe act or condition. It is the role of the scene safety officer to monitor activities and identify those unsafe acts and conditions.

Although a safety officer is vital to emergency operations risk management, such concerns begin long before arrival at the scene. Inspections of buildings throughout response districts, review of material safety data sheets in high-risk occupancies, and development of appropriate response plans are all examples of how risk management can be applied prior to an incident taking place. Considering this, it becomes even more obvious why the scene safety officer must have advance knowledge of conditions that might be encountered on the fireground as well as a full awareness of the role he or she is to play. It further indicates the need for the department to stipulate minimum qualifications and for the safety officer to be able to function within any version of the incident command system.

## emergency preparedness

In Chapter Three, we performed an exercise in which we identified potential loss exposures, then designed steps that we could take to lessen the vulnerability. In protecting firefighters from injuries, the process starts long before an alarm is received. Properly trained and qualified individuals, adherence to the latest safety standards, upgrading equipment inventories, and following a progressive preventive maintenance program are all part of emergency preparedness. The exercise in Chapter Three may have seemed to be an *administration* task, and in some cases that may be true. In reality, its purpose is to create a component of our risk management program that will be geared toward increasing safety during emergency responses. In many ways, being proactive means being prepared, and preparing for an emergency response is actually the first step toward managing risks if and when an incident occurs.

## emergency responses

Responding to an emergency call means getting to the scene quickly and safely. The consequent safety concerns can be planned, but they must also be flexible. Changing traffic conditions, road construction, detours, and other obstacles can all require you to modify your response protocols. Whenever such adaptations are required, you must keep safety considerations in mind.

Sometimes the complicating factor is known beforehand. If road construction in a given community is a concern, the department should have an opportunity to know about it in detail to plan alternate routes. Identifying situations that may compromise safety is another aspect of emergency risk management.

In addition, the department has an obligation to the community it serves as well as to its own employees. All individuals who are responsible for driving fire apparatus under emergency conditions should be properly trained and qualified. This can be a serious challenge for small departments and those in rural areas, where the driver of the apparatus is often the first person to arrive at the station—anyone who has the proper type of driver's license under local laws. Using improperly trained or unqualified drivers may present a serious liability exposure to the municipality or district if an accident occurs. Departments faced with this problem are obligated to send their drivers to an approved emergency vehicle operations course. These courses are designed so that the trainee must not only complete an instructional phase but must also demonstrate proficiency in practical skills. A department that doesn't have this type of program readily available should contact its insurance company, the state training authority, or another qualified instructor to make suitable arrangements. The safety officer, in conjuction with the training officer, should review the qualifications and licensure of personnel at least annually. Any member whose certification has expired, whose license is not of the proper class, or whose driving privileges have been suspended or revoked should be disallowed from operating fire vehicles under any circumstances until the appropriate requirements have been satisfied.

Just as certain requirements are imposed on motorists, personnel must also be capable of operating fire pumps and any other equipment to which they may be assigned. There are documented court cases in which responding fire companies failed to perform competently, resulting in additional loss to the property owner. The department must recognize its obligation to the community and ensure that it provides qualified personnel in all areas of operation. Clearly, additional training must be provided for individuals who operate fire apparatus under emergency conditions. Failure to do so could be so catastrophic, legally and otherwise.

## qualifications of the safety officer

The safety officer is an individual whose responsibilities carry great weight and whose actions affect everyone in the department and the community. It must be someone who is trained to an advanced level not only in fire prevention activities and specialized operations but also in areas of safety and human resource management. This officer will often be forced to make decisions that affect the outcomes of emergency operations and consequently the lives of civilians. To be successful in the fire department, the safety officer must possess a variety of personal attributes and professional qualifications. His or her personal traits should include:

- a mature approach when dealing with people under all sorts of circumstances,
- an ability to work autonomously while functioning within the incident command system,
- good communication and interpersonal skills, and
- honesty and integrity.

This officer's professional qualifications should include:

- knowledge of current safety standards to ensure that the department is in compliance in all phases of its operations;
- a working knowledge of all testing and diagnostic instruments used to test hazardous areas for toxic fumes, explosive gas levels, and other hazards;
- an ability to promote occupational health and safety within the organization;
- an ability to recognize unsafe conditions at emergency scenes, to predict potential hazards, and to implement appro-

priate corrective measures within the incident command system, and

- an ability to work with administrative officers, line officers, and fire department personnel at all levels to identify existing safety hazards and to recommend measures to eliminate those risks.

In terms of duties, this officer must also oversee:

- instructor certification to provide safety training for administrative officers, line officers, and other firefighters;
- the development of health and safety policies, procedures, and manuals with the assistance of other designated department members;
- identification of the causes of injuries that occur under both emergency and nonemergency conditions to recommend remedial action so that they are not repeated; and
- the rehabilitation of firefighters after accidents to ensure their satisfactory return to work, consistent with specific requirements of the job description.

Given the demands of the office, fire departments all too often appoint individuals without ever providing them with proper training. This is an injustice to the candidate (whose inadequacies will quickly be recognized within the organization) just as it is to those employees who will be turning to him for guidance and advice in critical moments. The fire department safety officer is a position that affects all areas of operation within the incident command system. It is the one position that has the authority, under certain conditions, to countermand orders issued by the incident commander. The need for proper training cannot be overemphasized. Without it, the safety officer will lose respect, and the department's efforts toward safety management will have been for naught.

## questions for discussion

(1) Has your department appointed a safety officer? Has this role been defined to include scene safety management as well as administrative duties?

(2) Has your department committed the time and money to ensure that the safety officer receives an appropriate amount of continuing education to keep abreast of changing safety standards in the fire service?

(3) Can the efforts of the safety officer be seen throughout the organization?

(4) What does your department need to do to improve its safety management program? How does the safety officer fit into these tasks?

# chapter eight: risk management and the quality management program

Your fire department risk management activities should actually be integrated into a much larger picture, that of Total Quality Management. TQM is a concept that originated in Japan following World War II. Early versions focused on improving the quality of manufactured goods. Since then, the concept has expanded into virtually all aspects of American business, including retail, health care, and service management, just to name a few. It has only been in recent years that the TQM concept has been recognized and implemented by emergency services. TQM looks beyond production issues, seeking to improve quality in all aspects of an organization's operations.

Traditionally, the fire service has always viewed itself (and civilians have viewed it) as an entity of government that provides a service in times of crisis. By the early 1990s, this perception had begun to change. Increasingly, fire departments see themselves as business entities. As a result, a major paradigm shift has occurred. What was once seen as an industry focused on slobbering water has now become an industry focused on the people it serves, intent on providing nothing less than top-quality service. As budgets continue to shrink, and administrators are forced to do more with less, many departments find themselves having to turn back to their communities for financial support. Quality management plays such an important role today because, without satisfied customers, fire departments stand less chance of garnering the public support they need.

Much as is the case with the risk management measures previously discussed, many fire departments are already actively pursuing quality management programs, often in a haphazard manner. With or without an organized quality management program, think about a few of the quality improvement programs that many departments have instituted in recent years. Whether mandated by regulatory agencies or designed by specific departments, all of these programs contain features of quality management. Some exemplary features include:

- the establishment of minimum training requirements for firefighters engaged in suppression and related activities,
- specialized service teams to broaden the scope of services provided by the department,
- implementation of customer satisfaction surveys to get feedback from the public on how the department performed and where room for improvement remains, and
- technological improvements designed to increase the efficiency and safety of fire department operations for all concerned.

These are just a few TQM examples. Your department has probably implemented tens, if not hundreds, of changes in recent years, all designed to improve the quality of services. Whether they have been instituted randomly or as part of an organized TQM program, all are germane to the concept of total quality management.

A good fire department TQM program may be expected to produce some easily recognizable benefits. Depending on your initial circumstances, these may include:

- increased productivity,
- increased professionalism,
- improved response times,
- appropriateness of services provided,
- improved financial management of the department,
- increased customer satisfaction,
- improved public support, and
- increased awareness and involvement within the department.

## designing a fire department TQM program

Any TQM project starts with two basic questions: What area of operations do we want to assess, and what data collection system do we have available?

To be successful at TQM, most fire departments must accept taking one step at a time. In urban areas, committees may be large enough to examine several operational areas at once, but this is the exception rather than the rule. In looking at an operation, decide what elements of it are to be studied. For example, if the department is studying its response times, then limit the focus of the initial study to response times only.

The next critical element is data collection. To turn data into useful information means that only data relevant to the study should be collected. If response times are at issue, then the data collected should be related to the influencing factors, such as actual times, travel distances, travel routes, time of day, traffic conditions, and the like. Avoid collecting irrelevant data, since it will only complicate and confuse the study, resulting in unreliable outcomes.

Once all of the data has been collected, the department can develop a model of the trend. This enables the department:

- to make a scientific decision as to whether a problem truly exists or whether it is only perceived;
- to compare its performance against any applicable standards;
- in the absence of standards, to develop its own goals or standards; and
- to develop plans to attain new goals or standards.

Once the data has been analyzed and the trends identified, the department can look at the underlying causes. Normally, these will be broken down into four categories:

- work practice controls, such as inaccurate or ineffective policies or procedures, or poor record-keeping systems or requirements;
- engineering controls, such as proper types and use of equipment;
- administrative controls, such as adequate staffing, proper scheduling, and appropriate training; and
- personal protective equipment.

In most cases, as the underlying causes become obvious, so do the probable cures. From its information, the department should be able to develop an appropriate course of action. This approach to TQM mimics the techniques inherent to the five-step risk management process. Finally, after implementing the corrective measures, the results must be monitored. Doing so means that the program is not only *total* but *continuous* as well. Once the department reaches its goals, it should establish new ones so that it is constantly striving to improve the quality of its services.

## risk management in the TQM process

When you think of the fundamental concepts of a fire department risk management program, you think in terms of loss prevention, safety management, and asset preservation. These are intimately related to providing quality services. The link between your risk management program and your TQM program will be very strong once both are established. Since the fire service as an industry accentuates self-critique, we frequently have the opportunity to assess quality and risk management measures. Every completed incident report gives us the opportunity to conduct such an assessment. Every firefighter or civilian casualty provides us with a learning opportunity that can result in improved services. Every complaint or concern allows us to form new insights about our operations. All of these demonstrate the essential relationship between risk management and quality management.

Let's consider a hypothetical situation that allows us to use fireground outcomes with risk management implications as a tool of TQM. Suppose that Engine Companies 3, 11, and 22, along with Truck No. 12 and Ambulance No. 10, respond to a residential fire in an older section of their community. The call is in an area where the water supply is known to be limited and access is often challenging because of narrow streets and residential parking. Engine 3, the first-arriving vehicle on the scene, advises dispatch that the structure is a one-story home of wood-frame construction, 40 feet × 50 feet, with an apparent fire in the northwest corner, in what appears to be a bedroom. Heavy black smoke and flames are visible through a window. Engine 3 also advises that all of the occupants have gathered in the front yard and that everyone is accounted for. Engine 3 starts an interior attack. Engine 11, the second-due vehicle on the scene, backs up toward Engine 3, drops a large-diameter hose, and proceeds to the nearest hydrant, located 250 feet away. Since the hydrant is relatively close to the fire building, Engine 11 ties the large-diameter hose directly to the hydrant, forgoing any relay pumping techniques. Attack personnel from Engine 3 enter the front door of the residence and proceed down an interior hallway to the fire room. Since the door to this room has been left open, they encounter heavy smoke conditions throughout the building and now have fire impinging on the hallway as well as the opposing bedroom. Heavy burning gases are seen traveling along the ceiling of the hallway. Using their attack lines, the firefighters quickly break up the burning gases but note that the moment they shut down their hoseline while attempting to advance on the fire, the gases reignite. Personnel from Engine 22 report to Engine 3 and pull a second line as a backup. This line is taken to the front door of the residence where it is charged, and a company is assigned to stand by as a rapid intervention team. Soon the attack crew notices that the water pressure in the attack line has decreased considerably. They have no radio link either with the pump operator of Engine 3 or with any of the command officers. Because of this, and for their own safety, they decide to retreat from the building. While they are exiting, a flashover occurs. The entire crew from Engine 3 exits the building safely, although the last one out receives second-degree burns to his ears and neck and is subsequently transported to the local hospital. Outside, the attack crew learns that it had exhausted the available water supply in the booster tank while trying to keep the fire in check in the hallway. Evidently there had been an inadequate water supply available from the hydrant to fill the large-diameter hose, and suppression operations were compromised as a result.

What are the quality management and risk management implications of such a scenario? Since none of us were on the scene of this imaginary fire, our mental images of it will be vastly different. Still, the quality and risk management implications will be similar. These may include:

- the failure to recognize a compromised water supply, resulting in aborted suppression operations and a firefighter injury;

• the failure to provide adequate radio capability, which prevented command from being able to order the crew out of the building when the water supply diminished; and

• the failure to recognize the need for a relay operation, thus limiting the use of available water, resulting in an injury to a firefighter.

Once you have identified the various implications, try to ascertain whether each is a quality management issue, a risk management issue, or a combination of both. You'll probably find that most of them fall into both categories. By this scenario, we can appreciate the essential correlation between quality management and risk management. Often they will seem similar, if not the same. Other times, each will have its own unique aspects. You probably noted:

• the failure to recognize a compromised water supply (quality managment issue), resulting in aborted suppression operations and a firefighter injury (risk management issue).

• the failure to provide adequate radio capability (quality management issue), which prevented command from being able to order the crew out of the building when the water supply diminished (risk management issue).

• the failure to recognize the need for a relay operation (quality management issue), thus limiting the use of available water, resulting in an injury to a firefighter (risk management issue).

Please note that we have only identified three implications. In the real world, you should be able to find many more. Even the amount of information given here should have been enough to stimulate thought in related areas—the appropriateness of personal protective equipment, for example. You can probably appreciate how quality management and risk management considerations are similar, yet different. Either program can work alone, but when they are combined, the department and the public reap double benefits (Figure 8-1).

Figure 8-1

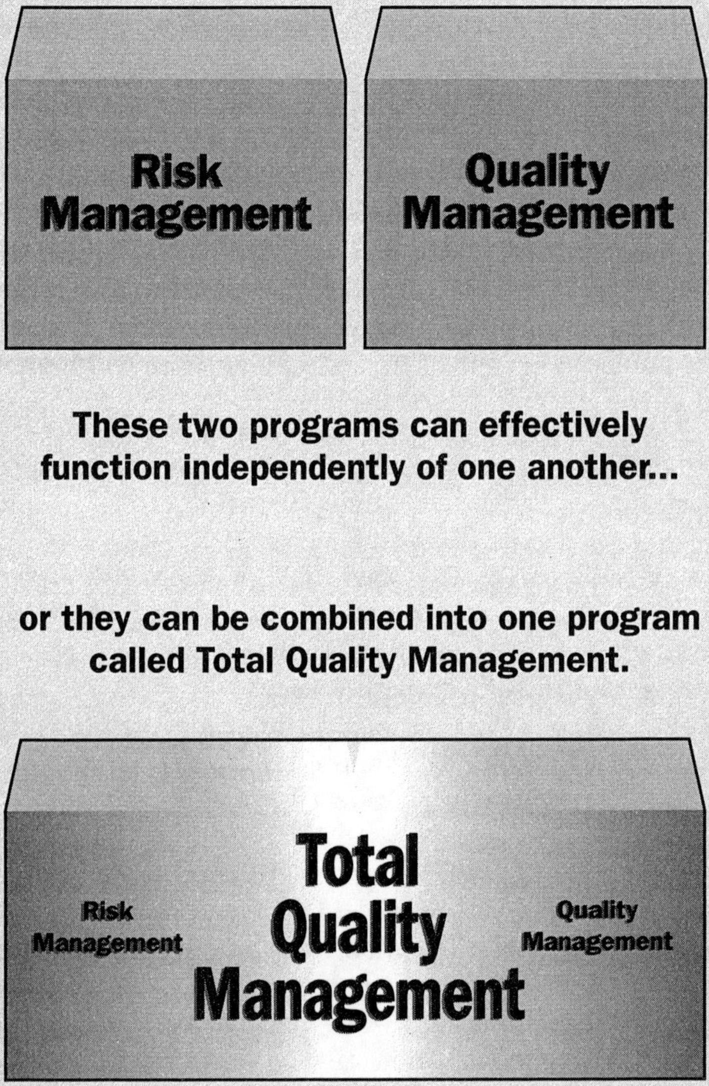

Risk Management

Quality Management

These two programs can effectively function independently of one another...

or they can be combined into one program called Total Quality Management.

Risk Management

Total Quality Management

Quality Management

Much like risk management, TQM is a concept whose time has arrived in the fire service. Administrators must understand the distinctions of both risk and quality management if they are to develop workable programs and mechanisms within their organizations. Once in place, these programs will touch every area of department operations.

## questions for discussion

(1) In what quality management activities does your department currently engage? How could they be made better? Are there additional areas of operation that could be improved through quality management measures?

(2) Does your department communicate quality management findings throughout the organization? How is that information used in day-to-day activities?

(3) How do your quality management and risk management programs interact? Is TQM a part of your department?

# chapter nine: putting it all together

By now, you should have a clear understanding of risk management in the fire service—the benefits that such a program offers as well as the hazards of not having one. The next step is to put such a program together. Done systematically, its design and implementation can go quite smoothly. Done incorrectly, the program can collapse, failing as quickly as it began.

The design and implementation process requires a strong commitment from the administrative staff of the department, starting with the top-ranking officer. This officer can in turn convey his or her sense of commitment throughout the department by means of a commitment statement. This will serve to empower those individuals who are involved in the program's design. It will also send a signal to other employees that the program is serious business, that the department is committed to it, and that everyone's cooperation will be expected. Let's look at an example of such a statement.

*"The purpose of a risk management program in the ____________ Fire Department is to ensure that the department is prepared to*

*handle potential losses such as accidents, injuries, and deaths, as well as equipment failures and other similar occurrences, in such a way that the department is able to recover and continue to operate. It is also intended to identify the most effective means to safeguard the public and to provide all services, both emergency and non-emergency, in the best way possible while keeping unavoidable losses at a minimum."*

The next step in the process is to designate an individual within the organization to serve as the department risk manager. This person, empowered by the fire chief or his designee, is responsible for the program's overall design, implementation, and operational management. In conjunction with these responsibilities, this person must also be given the authority necessary to accomplish the desired goals. In many departments, the person designated risk manager is the same individual who serves as safety officer. If this approach works well for your department, that's fine. If the safety officer, for whatever reason, is unable to accept the responsibilities of risk management, then appoint another member with comparable qualifications. No matter who is chosen, it is imperative that this person interact well with the safety officer, since the two will often find themselves teaming up on issues of risk and safety management.

In most departments, the person designated risk manager hasn't had the opportunity to receive a considerable amount of training in the subject. The department is obligated to ensure that the candidate be properly trained. Without it, the individual will fail and the program will quickly lose its credibility. Keep in mind that risk management is a relatively new field. Few people have already developed the expertise necessary to function as risk management professionals within the fire service. The person you appoint will need additional training and education or at least access to an outside consultant for guidance and direction.

Once you have chosen the best person to lead the program, the next step is to provide him or her with the resources essential to the task. This entails both finances and support personnel. Whether you use the safety committee or develop a risk management committee, your risk management coordinator will need people to help him develop, implement, and manage the program. The members of this committee must be dedicated to the program, recognizing and believing in its potential benefits.

Next, the committee members should develop their goals (see Chapter Five). These must be realistic and quantifiable. This is especially important in the early development of the program. Most people are creatures of habit and are resistant to change. Many will be skeptical about the program and will be waiting to see it fail. Those involved in instituting risk management must recognize this and develop their early objectives in such a way as to be almost geared toward success in a fail-safe way.

After setting the goals, the group can then begin applying the five-step risk management process as detailed in Chapter Three. From that point on, the committee should be self-perpetuating, the process ongoing.

For years, most of us in the fire service have focused on saving lives and protecting property. The 1990s have brought us a new breed of firefighter. While the fire service was once reserved for some of the toughest and bravest members of society, it has now become a cerebral entity as well.

The concept of risk management in the fire service was unheard of 10 years ago. Today, it is virtually an expectation. Unfortunately, it can also be a complicated concept to those who don't understand it. Training and education are essential toward overcoming these obstacles.

Despite any effort, losses will occasionally occur. When they do, don't allow them to be seen as indicators that your program is ineffective. The best of programs in the most progressive of departments can't prevent all losses. Departments must use their losses in a constructive way, learning from them to prevent them from recurring.

By now, you should have an understanding of the fundamental concepts of a fire department risk management program. Use this material, and remember that risk management activities never end. A well-organized risk management program will help your department in every area of operation—today, tomorrow, and well into the future.

# chapter ten: sexual harassment in the fire service

Up to this point, we have looked at issues of risk management in terms of developing a risk management program, as well as at the basic elements of such a program. We haven't yet looked at specific issues of risk management, since they normally vary from one organization to the next. Throughout the fire service, one major issue is of growing concern to administrators, district trustees, and municipal officials: the increasing presence of sexual harassment.

Sexual harassment cases in the fire service are growing in frequency and severity. With an increasing number of women in the fire service, it is an issue that departments must take seriously. If a department ignores it or takes an it-can't-happen-here attitude, at some point and in all likelihood it *will* happen.

To understand the idea of preventing sexual harassment in the fire service, we must first establish a couple of definitions. *Sexual harassment* means any unwelcome sexual advances, requests for sexual favors, and verbal or visual material or conduct of a sexual nature. *Unlawful sexual harassment* is defined as harassment that is (1) a condition of employment, (2) a consequence of employment, or (3) of such a level that it creates a hostile work environment.

There are two basic types of sexual harassment. The first is *quid pro quo*, meaning "this for that." Quid pro quo sexual harassment simply means "If you do this for me, I will do this for you." Based on our previous definition of unlawful sexual harassment, quid pro quo harassment becomes an unlawful form, since it seems to suggest a condition of employment. Even when implied in a humorous format, this type of sexual harassment can easily be misinterpreted and just as easily prosecuted.

The second type of sexual harassment is known as a *hostile work environment*. This is created whenever the conditions in the workplace become such that the employee fears or regrets going to work because of the fear and anguish of being sexually harassed. When such a condition exists, and the issues of sexual harassment cause this type of feeling in the employee, then the actions that cause those feelings are unlawful.

This can be a very sensitive area and one that is difficult to defend. What one party may see as humorous and lighthearted another party may see as harassing and offensive. It is up to the fire department administration to be constantly aware of such behavior and to prohibit it from occurring in the workplace. If management is aware that such conditions exist and allows them to continue, then managment in a sense creates its own hostile work environment.

Often a question is raised regarding the statute of limitations on sexual harassment claims. Normally, the limitation is one year from the date of the last event. Since this may vary from one jurisdiction to the next, your department should discuss this issue with an attorney or other legal representative if it has any concerns.

### applicability of laws

Most federal courts have adopted what is called the *reasonable woman standard*. This implies that the actionable conduct to which the female employee was subjected is opposite that of what a reasonable woman would accept in the workplace. Note that that creates a definitional difference between reasonable *woman* and reasonable *person*, since it has been documented that women are affected differently than men when it comes to dealing with issues of sexual harassment in the workplace. This applied standard forces the jury to "think like a woman." In so doing, an aura of empathy is created in every jury member regardless of sex or race.

Another interesting area involves same-sex sexual harassment. For some time now, same-sex harassment brings up the question of applicability of sexual harassment laws when the offender and the receiver are both of the same gender. In one case, a male employee filed a sexual harassment charge against another male employee after the alleged offender jokingly grabbed him in the groin. The male employee who was grabbed didn't see the humor in it and didn't receive an appropriate response from his employer when he complained. The case went to court, which ruled that it wasn't sexual harassment and instead classified it as a case of criminal assault. This ruling forces the fire service, where such actions may seem common, to revisit its position on what it will and will not allow.

## employer liability

Although the courts have heard many cases, the general consensus appears to be that there are four common areas of employer liability under federal law. These include:

(1) acts of superiors, managers, and agents;
(2) acts of employees;
(3) acts of nonemployees; and
(4) liability to those who are *not* harassed.

Each of these areas presents unique challenges to the fire service administrator when addressing the issue of sexual harassment in the department.

*Acts of superiors, managers, and agents.*

Until recently, the position of the courts has been that the employer was liable for the way employees were treated in the workplace. Although some recent decisions seem to reverse this approach, a prudent fire department risk management program would be focused on preventing actions that could test this theory. Recent court decisions have protected the employer from prosecution; however, the employer must take certain preventive actions, which may later become the basis of a defense argument. Such actions will be discussed later in this chapter.

*Acts of employees.*

If the employer has met the requirements of federal law, the employee who acts in a way that constitutes sexual harassment may actually find himself or herself in a position of individual liability. If the victim of such actions can prove a claim, damages may be awarded with the judgment going against the defendant as an individual. If such is the case, it is highly unlikely that any insurance protection provided by the employer would cover that individual. In a sense, the defendant is left bare and exposed, meaning that his or her personal assets may be in jeopardy. Fire service administrators must realize this possibility and make every effort to educate their employees about the personal risks that harassing behaviors create.

*Acts of nonemployees.*

One common argument suggests that the employer shouldn't or can't be held liable when a nonemployee sexually harasses a member of the department. Unfortunately, this may or may not be true. No matter who commits the offensive act, be it a department employee or nondepartment person, if the employer is aware that such activity is occurring in the workplace, then the department may still find itself in a position of liability. Previously we talked about the issue of a hostile work environment. In cases where the offender isn't employed by the fire department but is harassing employees in the workplace, the department is obligated to intercede on behalf of the employee. In doing so, the department demonstrates a good-faith effort to provide its employees with an environment that is free of sexual harassment. If the employee committing the harassment is employed by another company, then the fire department administration must contact the employer of the offender and advise it of what is taking place. By doing so, the fire department places the outside employer on notice of the actions of its employee, sending a strong signal that such behavior will not be tolerated. At the same time, the administrator must be certain to document all conversations, which may later prove that a prudent effort was made. The fire department should be able to show, once notified of a situation, that it took appropriate remedial action to eradicate the hostile work environment as well as to prevent future incidents. Appropriately managed, issues of sexual harassment caused by the actions of nonemployees of the fire department shouldn't become issues of fire department liability under federal law.

*Liability to those who are not harassed.*

This is a delicate issue—one that could become a major headache for the fire deparment. The concern that this creates can be summed up in one word: *favoritism*. Many companies and corporations have recognized the exposure that they inherit when personal relationships develop between supervisors and subordinates, and the fire service should show the same concern. If a supervisor and a subordinate are involved in *any* type of relationship—be it personal, sexual, or otherwise—and *if* the subordinate subsequently receives any type of promotion, salary adjustment, or any other type of work benefit that was not afforded to the other employees, then favoritism seems to have been a factor. Such situations, when allowed to exist, may put the department in a position of having to justify any actions involving those employees. If another employee is more qualified or deserving of the raise or promotion, he or she may very well file a liability claim against the employer, arguing that the other person received those benefits unfairly. This type of argument can create an unpleasant environment within the organization and is clearly one reason that organizations everywhere have adopted fraternization policies prohibiting relationships between supervisors and subordinates or between coworkers.

## case studies

Recent jury awards have sent a strong message to business, industry, and government that sexual harassment will not be tolerated in the workplace. While an employer may choose to defend such a case aggressively, in the end it will lose despite the legal outcome. For any case to be tried, the cost of defense can be astronomical. It is not surprising to see the cost of defending such claims exceeding $250,000, plus any damages that may be awarded by the jury.

The recent propensity of juries that hear such cases has been to award extremely large amounts in damages. In one case, a major national retailer was ordered to pay $50 million in damages after a male supervisor made comments about a female employee's breasts and buttocks. The comments were verbal only, and there was never any suggestion of physical contact. The retailer appealed the award, which was later reduced to $5 million. With or without the appeal, such an award sends a strong signal about the mindset of jurors when dealing with issues of sexual harassment. The empathy factor runs high, and the ability to defend and/or justify harassing actions is difficult.

In another case, a female employee reported to work wearing a blouse without a bra. Over the blouse, she wore a blazer type of jacket. Realizing that she was bra-less, her boss dropped two M&Ms into her coat pocket, then reached inside to get them, intentionally brushing her breasts. The female employee filed a complaint and the case went to court. The jury awarded her $7.1 million, which was later reduced to $3.5 million on appeal.

In another recent case of an even more unusual nature, a jury awarded $500,000 to a 14-year-old female after finding that school officials ignored her complaints about a classmate who was sexually harassing her. The jury award was broken down so that the school district had to pay 93 percent of the damages. The boy's family was responsible for $27,000, and the former principal owed $6,000.

Finally, a San Francisco court awarded a former female police officer $288,000 in damages stemming from her sexual harassment lawsuit against the former chief of police. The involved officer accused the former chief of sexually harassing her, in a form that was both physical and verbal, on numerous occasions between 1989 and 1993. She also claimed that she was demoted from being department spokesperson back down to patrol officer after refusing the former chief's advances. She has since quit the police department.

These cases share similarities, the most obvious being that damage awards are rising and that juries are making awards that reflect their empathy toward victims.

Fire departments must realize that, regardless of other applicable statutes that may suggest governmental immunity, they are bound to comply with the same laws on sexual harassment as any other organization or corporation and must meet all state and federal laws that deal with sexual harassment in the workplace.

## prevention: the best cure

A sexual harassment claim, much like any other liability claim against a fire department, is a no-win situation. No matter how a judge or jury may rule, the department ultimately loses. Publicity of the events leading up to the trial, the cost of defense, investigative costs, and negative impacts on morale all add up to substantial losses. The best way to win the battle against sexual harassment is to prevent it from occurring in the first place. For a department to prevent claims or allegations of sexual harassment in the workplace, it must:

- have an effective sexual harassment policy,
- train its employees,
- maintain an open-door policy,
- conduct harassment audits, and
- conduct appropriate and effective investigations of *all* complaints.

*Have an effective sexual harassment policy.*

An effective sexual harassment policy contains several simple components, all of which spell out the department's position on sexual harassment in the workplace. Such components include:

- a zero-tolerance policy, clearly stating that the organization will not tolerate sexual harassment in the workplace in any form.
- a description of the prohibited types of behavior, consistent with federal laws, so that employees are aware of what types of behavior are unacceptable.
- an explanation of an employee's right to report harassment without fear of retaliation. It is very important that all employees understand their rights under this policy and under federal law.
- a procedure whereby a victim doesn't need to confront his or her alleged harasser. This allows management to do its job without further complicating an already tense situation.
- an explanation of the employer's response, telling employees what actions and processes will take place once a complaint is filed.
- a prompt investigation. An allegation of sexual harassment is a serious complaint and deserves an immediate, formal response by management.
- provisions for immediate follow-up, including disciplinary action. If the findings of an investigation substantiate the

complaint, the department must take appropriate remedial or corrective action. In doing so, the department must remember that the offending employee still has his or her rights and that disciplinary action must be done confidentially. It is not a matter of public record.

*Train the employees.*

The policy on sexual harassment developed and adopted by the department must be shared by all employees. *Sharing* doesn't mean distributing the policy—it means that the department will provide education and training to its members so that they will understand the content of the policy, the types of behaviors that are disallowed, and all of the other conditions and parameters relevant to the issue.

Emphasize the zero-tolerance policy. Make sure that everyone understands that all employees throughout the heirarchy are governed by this policy and that there will be no exceptions. Also, explain the types of harassment and why they won't be tolerated. You can't assume that everyone who will be affected by this policy will necessarily understand exactly what is and what isn't tolerable behavior. A clear-cut explanation must be given to all. Explain the disciplinary process, too. Make sure that your employees understand how a complaint will be handled if one is filed. If yours is a union department, explain to them the role that union representation will play in the investigation, as well as the role of administration.

*Maintain an open-door policy.*

In the end, it will be in your best interest if you allow individuals with such concerns to visit you directly. For a person to step forward and report behavior that may constitute sexual harassment is in itself an act of courage. Usually someone does it on the chance that his or her working conditions and relations with coworkers will improve.

By maintaining an open-door policy, you send a strong signal to the employees that zero tolerance is not just a phrase used in policy but that it truly represents the administration's position. An open-door policy also allows you to stop a problem at the outset. This isn't to say that just having someone visit you is going to solve anything. Again, allowing people to come forward and speak to you directly sends a message to everyone in the department that you aren't going to tolerate harassing behavior. When employees realize that your stance on this issue is serious, there is a good chance that you can maintain a harassment-free environment.

Keep an open mind, but be practical. At times, you will no doubt receive complaints against individuals whom you may know personally. Some may have been friends for many years. Despite your relationship with an accused member, you must remain open-minded to the possibilities. If, for whatever reason, you don't feel that you can fairly handle a complaint, you must find another officer or subordinate who can remain fair and impartial throughout the investigatory process.

*Conduct harassment audits.*

This is a relatively new notion, and one that has seldom been applied in the fire service. In a previous chapter, we spoke about the appropriateness of the safety committee conducting safety inspections throughout a given department. The harassment audit is based on the same principle. Such an audit, performed at random, affords the administrator or officer an opportunity to observe areas within the organization that harassment is taking place. Such audits should be conducted on a regular, unannounced basis and should be performed in all stations and department facilities. These audits should be designed to include:

- visual audits, so as to identify posters, pictures, calendars, graffiti, or body art that may be offensive to individuals, may violate an individual's values, or may in fact lead to the presence of a hostile work environment.

• verbal issues, including suggestive or offensive comments made by one employee to another, or about another, whether the subject employee is physically present or not.

• physical issues, observing for behaviors in which one employee makes inappropriate physical contact with another, despite gender.

*Conduct appropriate and effective investigations of all complaints.* It is important to remember that your response to the employee's initial complaint may in fact predicate what that employee's next step will be. If you respond appropriately and professionally, the employee may be quite satisfied. When handled appropriately, the issue will be managed quietly and amicably within the organization. If you mismanage the complaint or fail to take it seriously, the employee may feel that he or she has no alternative but to turn to the courts. Once this occurs, you will find yourself wishing you had taken it more seriously in the first place.

When an employee approaches you with a complaint of harassment, you should first inform the complainant of his or her rights under the harassment policy of the department, plus any other applicable rules, such as grievance or employee advocacy policies. Assure the employee that you take harassment complaints seriously and that you will follow through on his or her complaint to an appropriate conclusion.

Next, interview the involved parties, starting with the complainant. This can be the most difficult part of the process, depending on the type of harassment being alleged. It is critical that the complainant describes to you exactly the type of behavior that has been demonstrated. If the form of harassment is physical, then you need to document exactly how, when, and where that contact took place. If the harassing behavior has been verbal, you need to have the complainant describe almost verbatim the comments that have been made. Realize that often you may be asking the complainant to say or describe things that are out of his or her character and thus embarrassing. It is up to you to put the complainant at ease and to make him or her realize how important this information is to any thorough investigation.

When you interview the complainant, ask for the names of others who have witnessed the harassing behaviors. You must interview these individuals as well. In doing so, be careful to avoid any comments or actions that could be construed as "finger pointing" toward the alleged offender. Question the witnesses about what behaviors they may have seen or heard; take all the necessary information, but show no bias.

Once you have interviewed the complainant and the witnesses, you'll probably have a feeling about the validity of the complaint based on the relative similarity of their stories. If justified, interview the alleged harasser. He or she also has a right to tell his or her side of the story. Be prepared for an *ignorant response*, meaning that the alleged harasser will in all likelihood deny any knowledge of the incident or the circumstances that you are referencing. When this occurs, allow that individual to continue speaking. More often than not, your silence will prompt additional comments. Often the harasser will talk himself or herself right into an admission of the scenario that you were investigating. Use silence to your advantage and let the person talk. As before, it is important that you not form any preliminary judgments or biases. At this point, all you are doing is collecting information. Once that has been done, you will have the opportunity to analyze it and make an informed decision as to the validity of the complaint. Until you reach that point, everything else is fact-finding.

Based on the information gathered, make your decision and communicate the results. Call the complainant into your office. Explain to him or her what you have done and what your findings are. If the complaint is valid, assure the complainant that you are taking appropriate corrective actions against the offender. Do not specifically state what actions you are taking. This may violate the offender's rights to privacy under various employment laws. If the complainant insists on knowing what actions you are taking, you can explain the offender's right to privacy, but don't expect the complainant to receive this information graciously. You must also be aware that disciplinary or corrective actions are not to be discussed or used as examples when training the rest of the department on sexual harassment issues. Using actual case studies within your own organization can be extremely dangerous, since ours is an environment where secrets are seldom kept. The moment you start referencing an incident within your own department, others receive confidential information from management. Thus, laws guaranteeing privacy and confidentiality are breached.

Sexual harassment in the fire service as an issue seemed to make its first appearance in the late 1980s, growing to alarming proportions in the 1990s. All indications are that this concern will remain with us on into the next decade. The professional fire service administrator, in conjunction with the personnel department or officer, as well as the risk management team, must be aware of the increasing risk presented by this phenomenon. All must be certain that they meet the intent of all state and federal employment laws; that their actions are consistent with the mission of the organization; and that all are diligent in their attempts to provide a work environment that is free of harassment, abusive treatment, and inappropriate action. Those departments that fail to recognize the risk and thus fail to manage it proactively may ultimately find themselves the subject of public scrutiny. They may very well have to deal with a financial loss, the loss of public trust, and even the demise of the good name of the organization.

**you are the judge**

Do any of these scenarios represent sexual harassment?

*Scenario No. 1:*

Robin works as a secretary in the Fire Prevention Bureau. Through the course of the workday, she is subjected to suggestive jokes, remarks, and questions about a woman's sexual desires. She is "accidentally" touched repeatedly. Is Robin the victim of sexual harassment?

*Scenario No. 2:*

Karen and Mark are assigned to the same truck company and end up spending a considerable amount of time together. Mark has asked Karen out on several occasions. She consistently rejects his offers, but Mark believes that persistence will win out, so he continues to ask regularly. Is he harassing Karen?

*Scenario No. 3:*

Linda is very attracted to the captain of her engine company, Dan. Since they are both single, she asks him over to her house for dinner. After a pleasant meal and a few too many drinks, they end up spending the night together. Could this be considered sexual harassment?

The behavior shown toward Robin in Scenario No. 1 clearly seems to constitute sexual harassment. Any repeated offensive or suggestive comments or inquiries about a person's sexual behavior are considered harassing, especially if the behavior continues even after the victim has asked the offender to stop. Further, any touching of intimate body parts, whether accidental or not, could be construed as sexual harassment. This could include stroking someone's hair, rubbing shoulders, putting your arm around a person, or similar actions. Here, it appears as if Robin is the victim of sexual harassment.

In Scenario No. 2, also, it would appear that Mark is harassing Karen. She has made it clear that she doesn't have any interest in him outside of the workplace. Mark's persistence in asking her out may constitute a hostile work environment for Karen, thus putting his behavior in the category of harassment.

Initially, it would appear that the behavior demonstrated in Scenario No. 3 doesn't constitute sexual harassment. The involved individuals are of consensual age, they were off duty at the time, and they spent the night together of their own free will. However, the fact remains that Dan is Linda's company officer. Although the present situation seems peaceful and amicable, differences could arise later that might result in ill feelings toward one another. At that point, it would be quite easy for Linda to file a harassment complaint, claiming that she felt forced to allow Dan to spend the night with her since he was her company officer. This is obviously a case in which there is no black-and-white answer. For Linda, it would probably be a difficult case to win. For Dan, as well as the rest of the department, it would probably be a no-win situation.

It is because of the potential of cases like these that more and more organizations have adopted policies prohibiting relationships between supervisors and subordinates outside of the workplace.

# appendix A: studies in case law

The following pages contain brief reviews of actual cases in which the fire department was the defendant.

The purpose of reviewing these cases is not to stir debate over the applicability of specific legal doctrines. Rather, these cases should be used as a learning tool. Look at each on its own merits, and identify how you as the department risk manager might have taken proactive steps to prevent such cases from ever arising in the first place.

The comments contained herein represent the opinions of the author, and in no way are they meant to be construed as legal opinion. Nor should any of the comments herein be construed as legal advice.

## case 1:
### *mismanagement of an incident*

At 10:30 a.m., the plaintiff in this case called the fire department in reference to an automobile fire. On arrival, the fire department found a vehicle heavily involved with fire in the driveway of the residence. Using a portable extinguisher, personnel attempted to put out the fire. Instead, they blew burning materials out of the vehicle and against the fiberglass door of the plaintiff's garage. The garage door subsequently ignited.

The fire department had not previously stretched any hoselines and did so only after the garage door was set afire. Once they stretched the hoselines, they ignored the request of the property owner to enter the garage to prevent further fire spread. They fought the fire entirely from the outside, causing extensive damage to valuable merchandise stored inside. The garage and its contents were subsequently destroyed.

The case was brought to court, with judgment in favor of the plaintiff.

## case 2:
### *delayed response*

On November 27, four members of a family died of smoke inhalation when fire destroyed their home. Relatives of the deceased brought a complaint of negligence against the city and the fire department.

The suit claimed that the fire department was called to the fire but that it was delayed in responding because the firefighters were absent from their regular duty station. Because of their absence, it was at least 15 minutes before the first companies arrived on the scene. The plaintiff argued that the response time should have been less than five minutes. When the firefighters arrived, some were noted to have an apparent odor of alcohol on their breath, and they were unable to fight the fire as trained professionals should.

The plaintiffs alleged that the timely response of the department, combined with a proper professional manner in fighting the fire, would have prevented the deaths of their family members. The fire department defended itself on the basis that the regular driver was sick and that no backup driver was available.

The state supreme court ruled that the fire department had a duty to respond immediately to the plaintiffs' call from the plaintiffs even though the driver was sick. The court ruled in favor of the plaintiffs.

## case 3:
### *injury during a fire drill*

The plaintiff in this case filed a suit against the city fire department following an incident on April 28. The department, through its Fire Prevention Bureau, was conducting a fire drill on the eighth floor of a downtown high-rise. Following the plan written by the city fire marshal's office, the plaintiff and fellow employees of the company where the drill was being held were instructed by the fire marshal to assemble in a small corridor near a bank of elevators. The fire marshal further told the plaintiff to stand next to a heavy fire door because the corridor was "overcrowded" as a result of the drill. During the drill, a coworker unexpectedly opened the fire door, causing the plaintiff to be knocked to the ground and injured.

The plaintiff alleged that the city, through the actions of its fire marshal, was directly responsible for the injuries she sustained because they were "uniquely aware of the danger into which they placed the plaintiff when they instructed her to stand in front of the fire door, and that they were in control of the plaintiff at the time she was injured, and therefore owed a special duty toward maintaining the plaintiff's safety."

The case went to trial, and judgment was in favor of the plaintiff.

## case 4:
### *accident involving apparatus*

A lawsuit was filed against the department and the city following a motor vehicle accident that occurred at 3:30

a.m. while a fire apparatus was responding to a reported structure fire.

En route to the scene, the driver of the apparatus was unable to stop in time to avoid broadsiding a car that had failed to stop at an intersection stop sign. On impact, the car flipped, killing one teenage occupant and injuring two others. The impact put the apparatus out of service.

During the investigation following the accident, the driver of the apparatus admitted to the police that he had had three cans of beer 12 hours earlier while he was off duty. Although no criminal charges were filed, when the family members of the deceased teenager learned that the apparatus driver had admitted to drinking off duty, they immediately filed suit claiming that he had been intoxicated and careless in his operation of the fire truck. The city was named as a codefendant for allowing the driver to operate the fire truck after he had been drinking.

This case was settled out of court.

## case 5:
### *accident involving apparatus*

On the morning of August 4, a plaintiff sustained personal injuries when the motorcycle he was driving struck a fire truck at a city intersection. The plaintiff was in the left passing lane, and the pumper was proceeding eastbound, responding to an emergency call. The accident occurred when the fire truck proceeded through the intersection against a red light and at a speed of 10 to 15 miles per hour. The plaintiff, approaching with a green light in his favor, collided with the rear wheel of the fire truck when he unsuccessfully attempted to stop his motorcycle. The plaintiff claimed that he had seen the fire truck before reaching the intersection and that he had engaged his brake. The plaintiff testified that he hadn't heard any horns, sirens, or other warning sounds. The evidence at trial revealed many issues of dispute, with accounts of the witnesses differing sharply. There was conflicting testimony about whether the fire truck driver had looked into the intersection before entering it; whether the fire truck driver had accelerated or decelerated once the truck was in the intersection, and whether the emergency warning devices had ever been activated. The driver of the fire truck acknowledged at trial that he knew he had a statutory duty to check the color of the light at the intersection before proceeding and that that was implicit in the level of care that he was obligated to satisfy, even within the rules of emergency right-of-way.

After presentation of evidence by both sides, the jury deliberated and returned a verdict in favor of the plaintiff.

## case 6:
### *slow response to a medical emergency*

On March 1, a female who felt weak and in need of medical attention called 911 and requested an ambulance. In taking the 911 call, the emergency dispatcher said that he would send an ambulance immediately. When no ambulance arrived for at least 10 to 15 minutes after the initial call, a friend of the patient made a second call and was again told that an ambulance was on the way. No ambulance arrived, and they made a third call, approximately 20 to 30 minutes after the first attempt. Firefighters eventually arrived "30 or more minutes" after the first call had been made.

The only equipment that the firefighters brought was an oxygen mask and a mouth-to-mouth resuscitation piece. According to the plaintiff, the firefighters walked slowly to the house despite calls from family members to hurry. Family members at the scene testified that the firefighters acted "casually and slowly" in approaching the patient, and that one firefighter used the rubber antenna of his radio to lift the victim's eyelid rather than touch her. The same firefighter made an undirected comment to "move her to the living room," which the family members did. The firefighters then administered

CPR. Sometime after the firefighters arrived, EMTs from the trauma unit also arrived and performed advanced life support techniques. They rushed the victim to the hospital, where she died. The doctor in the emergency room told the family that they could have saved her if they had gotten her there a few minutes sooner. The family of the deceased filed a negligence action against the fire department and the city. The trial court dismissed the complaint. The plaintiff appealed, and the appellate court reversed the trial court's decision, noting that a "special relationship" arose at the point when any one of the three calls was made, as well as when the firefighting crew arrived at the scene for the purpose of providing medical assistance. The case was remanded back to the trial courts for a hearing. The outcome was still undecided as of the date of this publication.

### assessing these cases

What elements might have been included in a fire department risk management program that could have lessened the chances of such cases ever arising? Could a proactive risk management program have prevented these losses? Why or why not?

In partial answer, each of these cases, while unique, represents an instance where a strong risk management program *could* have helped to avoid a loss. As previously stated in this book, there is no guarantee that a good risk management program will prevent professional liability claims. However, as the fire department risk manager becomes more experienced and astute, he or she will no doubt begin to recognize circumstances that could be actionable. It is then the risk manager's task to work with department administrators to develop policies designed to maintain a high level of service, while at the same time lessening the department's exposure to liability.

It should be noted that, in many of the cases referenced above, the department and/or the municipality or district attempted to use the doctrine of governmental immunity as the basis of its defense. In each instance, one or more courts ruled that, by the nature of the case, the governmental immunity defense plea was not applicable and that the department could be held accountable for its actions.

For more information regarding the doctrine of governmental immunity, or to better acclimate yourself with the laws in your jurisdiction, I recommend that you meet with your legal counsel in advance rather than wait for a claim to be filed.

Finally, as mentioned throughout this text, risk management should be proactive rather than reactive. All too often, ways to cover a loss exposure are only thought of after an actionable event has occurred. Lawsuits against fire departments and protection districts are uncommon. That is not to say, however, that we as an industry can sit back and ignore the obligations that we have. Nor should the fear of claims become the basis on which we make our decisions. If the fire department administrator fully appreciates the concept of risk management, appoints a competent risk management coordinator or liaison, and follows prudent risk management guidelines, then that administrator should be able to manage the department and make decisions based on what is right rather than "what might happen if it goes to court."

# appendix B: common departmental loss exposures

The following list of common loss exposures* is provided to help you develop your risk management program. The 12 categories are based on the areas of operation of a typical fire department as described in Chapter Three. As you review these potential exposures, examine your own department's practices and add to the list as necessary.

## administration

- Failure to establish a level of service.
- Failure to document critical decisions.
- Failure to follow federal hiring and promotion guidelines.
- Failure to maintain required records and documentation.
- Failure to follow established policies.

## personnel

- Noncompliance with federal hiring and promotion practices.
- Lack of enforcement of training standards.

---

* Some of the material in this appendix is reprinted by permission of the Justice Institute of British Columbia Fire Academy.

- Poor equipment.
- Lack of adequate staff.
- Poor physical conditioning.

**communications**

- Dispatching to the wrong address.
- Failure to advise of possible delays.
- Lack of proper training for communications personnel.
- Lack of proper equipment maintenance.
- Lack of backup or redundancy system.

**in-station activities**

- Slips and falls.
- Poor housekeeping habits.
- Theft and vandalism.
- Lack of supervision of contractors and volunteers.
- Equipment failures.

**public education**

- Failure to establish a public education program, if required.
- Providing incorrect or incomplete information.
- Failure to provide *qualified* speakers and instructors.
- Injuries to people attending or participating in public education programs.
- Failure to include warnings on hazards associated with programs.

**fire prevention and inspections**

- Failure to enforce codes fairly or uniformly.
- Failure to identify code violations.
- Lack of follow-up after initial inspections.
- Failure to document findings of inspections.
- Providing inaccurate or unqualified advice or information.

**emergency vehicle operations**

- Failure to ensure that all drivers are properly licensed.
- Inadequate insurance coverage on vehicles and operators.

- Failure to obey traffic laws.
- Injuries to firefighters due to falls, sudden stops, and other mishaps.
- Inadequate vehicle maintenance.

**rescue**

- Failure to attempt rescue.
- Lack of proper tools and equipment.
- Tool or equipment failure.
- Lack of properly trained personnel.
- Allowing third-party interference or participation.

**EMS operations**

- Medical malpractice.
- Failure to supervise provisional personnel.
- Medical transport to inappropriate facilities.
- Lack of proper type or amount of insurance.
- Improperly licensed personnel.

**fire suppression**

- Failure to procure an adequate water supply.
- Excessive water damage.
- Inadequate manpower on the scene.
- Inadequately trained personnel.
- Equipment failures.

**overhaul**

- Incomplete overhaul that results in a rekindle.
- Excessive damage caused during overhaul operations.
- Failure to perform necessary overhaul.
- Failure to notify the owner or occupant of hazards.
- Injuries to firefighters during overhaul.

**salvage**

- Damage to property.
- Injury to bystanders.

- Accusations of theft by fire department personnel.
- Failure to secure items removed from buildings.
- Failure to attempt or perform salvage operations.